Introduction to Telephones and Telephone Systems

Third Edition

For a complete listing of the *Artech House Telecommunications Library*, turn to the back of this book.

Introduction to Telephones and Telephone Systems

Third Edition

A. Michael Noll

Annenberg School for Communication
University of Southern California

Artech House
Boston • London

Library of Congress Cataloging-in-Publication Data
Noll, A. Michael
    Introduction to telephones and telephone systems / A. Michael Noll.—3rd ed.
    p.   cm.—(Artech House telecommunications library)
    Rev. ed. of: Introduction to telephones & telephone systems. 2nd ed. c1991
      Includes bibliographical references and index.
      ISBN 0-58053-000-1 (alk. paper)
      1. Telephone.   2. Telephone systems.   I. Noll, A. Michael.
    Introduction to telephones & telephone systems. II. Title.   III. Series.
    TK6162.N65   1999
    621.385—dc21   99-10309
                                                                              CIP

British Library Cataloguing in Publication Data
Noll, A. Michael
    Introduction to telephones and telephone systems.—3rd.ed.
    —(Artech House telecommunications library)
    1. Telephone systems   2. Telephone
    I. Title
    621.3'85
    ISBN 1-58053-000-1

Cover and text design by Darrell Judd

International Standard Book Number: 0-58053-000-1
Library of Congress Catalog Card Number: 99-10309

10 9 8 7 6 5 4 3 2

Contents

Preface to the First Edition

T HE TELEPHONE surely must be considered one of the most marvelous
inventions of the communications era. Physical distance is con-
quered instantly, and any telephone in the world can be reached
through a vast communication network that spans oceans and continents.
The form of communication is both natural and unique, namely, human
speech.

Telephone communication can bring joy or convey sorrow. Simply
by lifting the telephone handset and dialing a few numbers, we can "reach
out and touch someone," as promoted in AT&T's television commercials
a few years ago. From its earliest days, the telephone was an instant success. There is no reason to doubt that this success will be diminished now
or in the foreseeable future.

The technology of telephone communication is quite simple and
straightforward, but the system that makes telephony possible and
affordable is quite vast, involving a variety of technologies that have
appeared over the telephone's hundred-year lifetime. It is a complex

system in its size and scope, yet the basic principles of the technology are understandable with a little study.

The size and importance of telephone communication justifies the need for educated managers and laypersons to understand its workings. Furthermore, the recent restructuring of the telephone industry in the United States—an event that seems destined to repeat itself in other countries—makes it even more essential that many people understand the workings of telephone technology.

The material presented in this book is intended for people without any detailed engineering or technical background. Some of the material does assume a knowledge of basic electricity and electronics, though even these concepts are defined in the Glossary at the end of the book. The reader who desires a deeper knowledge of basic communication electronics is referred to the author's *Introduction to Telecommunication Electronics* (Artech House, 1987). However, most of the material should be very understandable even without any knowledge of electricity and electronics. The terms and concepts essential to understanding the material are explained in the text. If not, then the reader can usually read on without any loss of understanding of the broader concepts.

The material in this book is based on graduate-level courses taught by the author at the Annenberg School for Communication at the University of Southern California and also at the Interactive Telecommunications Program at New York University.

Four major areas combine to make telephone communication possible. The first area deals with the telephone instrument itself and is called station apparatus. The second area deals with the various transmission media that carry communication signals over long and short distances. The third area is involved with the techniques and technology for switching communication circuits and paths so that any desired telephone can be reached. The fourth area, signaling, involves the methods for controlling the operation of the telephone network. Each of these four areas is treated separately in following chapters. We should always remember that all of these areas work together as a single system to create a telephone communications network.

The systemic aspect of telephone communication is quite apparent in the newest telephone communication service—namely, mobile cellular telephone service. Here, mobile units work with a multitude of base stations transmitting two-way radio signals. The operation of the whole system is controlled by a centrally located computer, which also controls access to the conventional public switched telephone network.

Digital switching and transmission technologies have had a great effect on telephone communication. The digitalization of voice telephony would appear to facilitate a natural migration to an all-digital communication system in which voice telephony and data communication are combined together. Thus, a small amount of material on the basic principles of data networks is included in the book to present a total system perspective of telephone communication.

For nearly a century, telephone communication in the United States was synonymous with AT&T and the Bell System. However, the advent of competition in telecommunication and the restructuring of the Bell System have revolutionized the telecommunication industry. To give an historical perspective to these events, the history of the Bell System is reviewed. An overview including some personal perspectives on these events is presented also. Technology has greatly shaped telecommunication, but the fragmentation of the Bell System was far more cataclysmic than ever caused by revolutions in technology.

Dr. Martin C. J. Elton of NYU's Interactive Telecommunications Program encouraged me to try my hand at teaching, and this gave me the opportunity to teach the course that produced the course notes that produced this book. Dean Peter Clarke of the Annenberg School for Communication at USC created an environment to support such scholarly endeavors as writing this book. I appreciate their contribution. In addition, I wish to acknowledge the influence during my early career at Bell Labs of Dr. John R. Pierce, Dr. Edward E. David, Jr., Dr. Peter B. Denes, Dr. Manfred R. Schroeder, and Dr. Max V. Mathews. They all helped me appreciate the importance of explaining science and technology to nontechnical people in an understandable fashion.

A. Michael Noll
Los Angeles, CA
March 1985

Preface to the Second Edition

IN THE FIVE SHORT YEARS since the first edition of this book first appeared, a lot has happened to the technology of telecommunication. All new switching machines are digital. Optical fiber forms the backbone of nearly all long-distance communication networks. Cellular telephone service has experienced phenomenal growth. Analog frequency-division multiplexing has been completely replaced by digital time-division multiplexing. Facsimile machines have become as ubiquitous as the telephone in most businesses. This rapid pace of technology obsoleted the first edition of this book and created the need for a revised and updated second edition. Much of the original material has remained, and new material has been added where appropriate.

The writing of this second edition was delayed for two years while I worked with Dr. John R. Pierce in revising his book *Signals: The Science of Telecommunication* (Scientific American Library, 1990). This project gave me an opportunity to reflect on my work with such telecommunication services as picturephone, video teleconferencing, and videotex. These reflections became the basis for a new chapter on services in this second edition.

Karen Ann Kurlander and Steve Fleischer at Bell Atlantic Mobile Systems reviewed the chapter on cellular telephone service. Various sections of the switching chapter were reviewed by Peter Settles and Carl Griffith of Northern Telecom and by Richard Q. Hofacker, Jr. and his colleagues at AT&T Bell Labs. I thank all these people for their help and for correcting many of my mistakes and misunderstandings. I also thank my editor Dennis Ricci for his assistance in producing both editions of this book. This second edition was written during my sabbatical from the Annenberg School for Communication, and I appreciate the support of the school during this project.

The writing of this second edition created an opportunity to improve on the quality of the figures; they were redrawn from my hand drawings by Richard Cook at the Annenberg School using a Macintosh computer. My publisher also worked on improving the overall appearance of the book. We hope that you are pleased with the result of our efforts.

A. Michael Noll
Stirling, NJ
December 1990

Preface to the Third Edition

WHILE TEACHING my graduate-level course at the Annenberg School for Communication at the University of Southern California this past spring, I knew that the second edition of this book needed to be updated. Much had happened in telecommunication during the 1990s. Wireless had become a tremendous success, including the development of a host of new digital approaches. Optical fiber had continued its revolution of the world of transmission. In one decade, nearly as many undersea cable systems had been placed under the Atlantic as in the entire four decades of undersea modern cable systems. Computer-controlled switching continued to shape the network. New demands for packet data are being made. Meanwhile, the very structure of the telecommunication industry has been changing, along with major new legislation that was passed in 1996. I was not surprised when my editor at Artech House, Mark Walsh, called to suggest it was time for a third edition to the book. I had already come to that same conclusion myself.

A number of colleagues reviewed material for this edition of the book, and I thank them for their assistance. I particularly acknowledge the following people who reviewed various sections of the manuscript: Skip Staples of Nortel Networks; Gary Schober of Berkeley Varitronics Systems; Tracy Scholle and Michael Kuehn of GTE Airfone Incorporated; George Huensch, Tom Bowers, and Meyer Zola of Lucent Technologies. Joseph J. Horzepa of Horizon Consultants reviewed the entire manuscript to help me with the cohesiveness of its content and also to improve its written style. Joseph W. Lechleider, retired from Bellcore, explained various technical and other material to me regarding wireless.

The third edition of this book was researched and written during the summer and fall of 1998 while I was on vacation and a sabbatical from the Annenberg School.

A. Michael Noll
Stirling, NJ
October 1998

1

Introduction

The key to the information age

Since the 1970s, communication technology has made great advances, offering capabilities that were unthinkable a decade earlier. The bandwidths possible with the new transmission technologies are vast and virtually unlimited, offering the ability to carry large numbers of voice, data, and video signals. New electronic switching systems are reliable and small in size compared to the electromechanical systems of the past, and the computers that control them offer great flexibility in terms of new programmable services. Simultaneous with these technological advances, the whole telecommunication industry continues to undergo dramatic restructuring.

It is easy to be swept away with an uncontrolled enthusiasm for the new communication technologies and the many new services that they could make possible. An optical fiber to every home could bring a tremendous variety of television programs on demand. All the world's

1

information could be delivered to every home in the form of information frames displayed on the home television set. The capabilities of the new transmission and switching technologies could make two-way video telephones feasible. The home itself could be controlled by a home computer that could be accessed remotely for monitoring purposes. Banking could be conducted electronically by telecommunication. Many employees could work at home, using telecommunication technologies to telecommute to meetings. Business meetings with geographically dispersed groups of people could be conducted by two-way video teleconferencing, thereby eliminating the need for physical travel. Television could become three-dimensional in high definition on large, thin displays that would fill the living room with images and sound. Voice, image, data, and video communication could all come together along with computers and telecommunication to create a single, integrated system. And all this could be only the beginning!

This utopian view of communication technology and all it could do is cloaked in a sense of euphoria that makes it difficult to separate hyperbole from reality. Decades ago, videotex promised instant access to computerized data banks, but the consumer acceptance of it was very disappointing. Today, the Internet offers access to the World Wide Web and is routinely used to send e-mail and to access databases. Two-way video teleconferencing continues to excite much interest, but the idea is quite old and its actual use is still small. The picturephone was tried decades ago and failed; most consumers simply do not want to be seen while speaking on the telephone.

Yet, there is one communication service that is over 100 years old and attracts little euphoria or excitement. We simply take it for granted that it is available in nearly every home and workplace in the nation, and in the world. We use this service eight times per day, on the average. With this service, emergency aid can be quickly summoned, airline reservations made, goods ordered and purchased, and funds transferred. No special skills are needed to use this service. This ubiquitous service that totally penetrates our lives is the telephone. Telephone service is truly synonymous with communication, but is so taken for granted that we frequently forget its paramount position in our lives and society.

Telecommunication is the key component of the Information Age and the services that form the basis of our information economy, and the data clearly supports this claim, as shown in Table 1.1. Telecommunications revenue for the United States in 1995 was $269 billion—far more than the $78 billion for outlays for personal computers; $82 billion for newspapers, books, and periodicals; and $133 billion for the cinema and television industry, including set sales, advertising receipts, video tape, and CATV. In terms of telecommunication traffic, there is no doubt that voice telephone service is the major contributor, although data traffic is

Table 1.1
Communication Industry Revenue in the United States for 1995*

Segment	1995 Revenue (in $ billion)
Telecommunication	$269
Telephone service	$178
Equipment sales	43
Cinema and television	$133
Motion pictures	$ 58
Television	41
CATV	34
Publishing	$ 82
Newspapers	$ 38
Periodicals	24
Books	20
Audio	$ 33
Radio stations	$ 9
Recorded media	12
Equipment	12
Personal computer	$ 78
Hardware	$ 68
Software	10
Physical delivery	$ 78
U.S. Postal Service	$ 48
United Parcel Service	21
Federal Express	9

* The data is summarized from *Stat. Abs. 1997*. All revenues are in billion dollars.

an important aspect of computerized information services and cannot be neglected.

The main thrust of this book is the telephone and the switched network, and the technology that makes telephone service possible. The information about the workings and evolution of this technology is mainly focused on developments in the United States. However, the basic principles apply universally, and hence the descriptions of basic technology should be useful to those interested in international telephony.

The vision

The telephone is generally acknowledged to have been invented by Alexander Graham Bell in 1876 with the assistance of his colleague Thomas A. Watson. Bell had the vision to see a world wired together by his invention. After a decade or so of intense competition, the Bell companies emerged as the monopoly supplier of telephone service in the United States. Theodore N. Vail founded the Bell System as a vertically integrated monopoly to supply telephone service. Nearly a century later, as a result of the drastic restructuring of the Bell System that occurred on January 1, 1984, competition returned to the provision of telephone service. In many ways, it was technology that led to the breakup of the Bell System.

Regulation

The provision of telephone service is subject to regulation on a number of governmental levels. Interstate long-distance service is regulated by the Federal Communications Commission (FCC). Intrastate toll service and local service are regulated by state public utility commissions, and possibly county and city commissions in some circumstances. As a result of the divestiture of the Bell local telephone companies from AT&T, the U.S. District Court, the U.S. Congress, and the U.S. Department of Justice now all also have a voice in telecommunication policy and regulation. From a policy perspective, telecommunications has many players and can appear very confusing and contradictory at times. The last chapter attempts to give some perspective on divestiture and telecommunication policy.

The consumer shapes the future

Telephone communication has not stood still during its century of existence, and new services and new uses of the public switched network continue to emerge. Some of these new services, such as video teleconferencing and videotex, are technologically most impressive, but consumers have expressed little interest in many of them. Other new services, such as mobile cellular telephony, are also technologically impressive, and consumers have responded enthusiastically to them. Technology indeed suggests many new services, but ultimately it is the consumer who determines success and failure in the marketplace. Later in this book, we will examine many of these new services and the apparent reasons for their success and failure.

System elements

The telephone system is a network, involving telephones, various transmission media, switching, and signaling to control it all. A network perspective of telephone service is covered in Chapter 2 as a general overview that should always be kept in mind as later chapters describe individual aspects of the provision of telephone service.

The technology of a telephone communication system consists of four major system elements. The first element is the telephone instrument itself, along with any other form of apparatus that is attached at the telephone station, as described in Chapter 3. The second element, transmission, is involved with the various media used to transmit a telephone signal over distance. In this book, transmission is covered in Chapters 4 and 5. The third element, switching, concerns the various ways in which one telephone circuit is connected to another telephone circuit. Switching is described in Chapters 6 and 7. The last element, signaling, is involved with how the telephone network is controlled and instructed to make an actual connection and is covered in Chapter 8. The technical

content of this book thus is organized into four main themes: station apparatus, transmission, switching, and signaling.

The provision of telephone service involves a system in which all the
elements must work together. The system is the telephone network.
Thus, the next chapter reviews the overall workings and basic principles
of the network that provides telephone service.

2

Networks: A Systems Perspective

Introduction

The ease with which any telephone can reach any other telephone is truly a marvel and involves a ubiquitous network, comprised of many networks at various levels and with differing geographic coverage. At the broadest conceptual level, telephones, facsimile machines, and personal computers all use and share the same network, as shown in Figure 2.1, which can carry signals representing voice, data, and image information.

A number of technologies work together to create a network. Voice and data signals are switched so that they reach only the intended destination. Signals are carried over distances over a variety of transmission media. The entire network of networks that facilitates end-to-end telecommunication is itself controlled by various signals that tell the

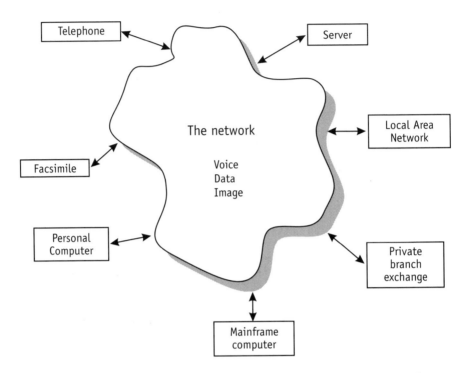

Figure 2.1   Telecommunication service is possible only because a network allows a wide variety of devices to communicate. Telephones, facsimile machines, personal computers, mainframe computers, and local area data networks can all share the network to communicate over distance.

switching machines what connections to perform and that assign transmission channels.

There are considerable equipment and facilities "beyond the dial" when a telephone call is made. The telephone instrument, along with a network of switched transmission circuits, enable any telephone to be quickly connected to any other telephone anywhere on planet Earth. This switched network is an extensive facility that allows the public to conduct voice conversations over almost any distance. The public switched telephone network not only connects telephones for voice conversations but can connect a computer terminal to a distant electronic data bank, a facsimile machine to another facsimile machine to transmit a picture or a letter across oceans. The same network can carry the sound of an ill heart to a specialist many miles away for sophisticated computer analysis and diagnosis. The telephone system and the public switched network are flexible resources that serve many communication purposes. It is here today, available for our use day in and day out, at our beck and call.

Historically, voice, facsimile, and local data traffic from personal computers has been carried over the telephone network, and high-speed data traffic has been carried over separate data networks, although these networks might share common transmission facilities. Today, these different kinds of traffic are converging, and the past boundaries between voice and data are dissolving.

The network must operate as an integrated system. This chapter describes the operation of the telephone network from this perspective of an integrated system. Later chapters will describe in detail the specifics of station apparatus, transmission, switching, and signaling.

Transmission, switching, and signaling

Signals are transmitted over physical distance over a wide variety of transmission media. Copper wire is used in the form of insulated wire pairs and also in the form of an outer conductor located coaxially around an inner conductor—what is called coaxial cable. Light waves are guided through single strands of optical fiber. Radio waves travel between towers located along the surface of the earth and also from the earth to communication satellites located high above the equator in geostationary orbit.

Transmission media are costly to install across continents and under oceans, and hence a medium must be shared by many users and their signals. Many signals are combined together through multiplexing. Multiplexing can be accomplished by giving each signal its unique band of frequencies—a technique called frequency-division multiplexing—or by converting the signal to a digital form and then assigning unique positions in a time sequence to each signal—a technique called time-division multiplexing.

One user of a communication network needs to communicate solely with another user. Thus, the communication signal produced by the user must be transmitted and connected only to that other user. This one-on-one nature of telephonic and most data communication makes it uniquely different from the broadcast nature of such mass media as television and broadcast radio. The connection of the signals of one user to those of another is accomplished by switching the signal as it travels from one location to another.

Initially, telephone calls were switched by physically connecting together wires—or circuits—to create a direct electrical connection from one phone to another that would be maintained for the duration of the call. This type of switching is called circuit switching. Today, most switching in the telephone network is accomplished by connecting the appropriate digital bits of one circuit to another—a form of virtual circuit switching since a virtual connection is maintained for the duration of the call.

Much data traffic consists of short bursts. It is only necessary to get the short burst of data from the source to the destination, and there is no need to maintain a complete connection for the duration of the data session. A short burst of data can be identified as to its destination and source and then switched gradually along a network as it travels toward its destination. This type of data switching is called packet switching and forms the basis for today's Internet. Packet switching and data communication are covered in more detail in a later chapter, although the major emphasis of this book remains the telephone network. Voice telephony and data might become integrated together in the future with the development of packet-switched networks that switch voice in the form of short fixed-length packets—a form of switching called asynchronous transport mode (ATM).

Telephone networks perform the switching and transmission of signals under orders and instructions issued over a control network, as shown in Figure 2.2. The switching machines in a network need to be instructed as to which connections to make, users need to be able to instruct the network about the identity of the party being called, and transmission links need to be identified as to which circuits are idle or in use. This aspect of control of the telephone network is called signaling. Considerable advances have been made in signaling, and many of the features that are associated with "intelligent" networks are made possible by modern signaling technology. The actual switching and transmission of customer traffic is performed over a transport network. Transport networks can be physical facilities.

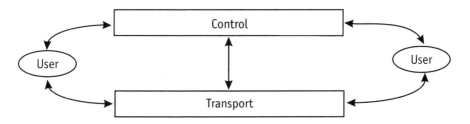

Figure 2.2   The transport of signals is under the orders carried over a control network. The issuing of orders used to control the telephone network is called signaling.

A network of networks

In the broadest sense, the term network means any means for connecting people with each other. The telephone network connects one telephone line to another. Telephones, facsimile machines, and modems can be connected to telephone lines thereby facilitating telecommunication between people by speech or image and between people and computers. Since the term "network" means any means of connection, the telephone wires in our home form an intrapremises network, the wires from our home to the nearby telephone-company switching office form another network, and the long-distance facilities of our long-distance carrier form yet another network. The telephone network thus is a network of networks, as depicted in Figure 2.3. All these separate networks must work seamlessly together when we use the telephone network, in the broadest sense, to connect with another person.

Some of the networks are determined by the geography of technology, such as the wires within a home or an office building. The local pairs of copper wire that connect homes to the telephone company's serving local office form a local network shaped by the local architecture of the technology. Long-distance networks span the continent and are also shaped and defined by geography. Other networks are defined mostly by policy considerations determined by government agencies. Telephone service within the United States was carved in 1984 into a number of local areas that are served by the local telephone companies. The local service areas of the telephone system are called local access and transport areas (LATAs). Calls between LATAs are referred to as inter-LATA calls.

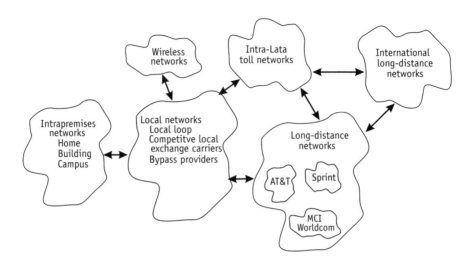

Figure 2.3   The network is comprised of many separate networks at the intrapremises, local, wireless, intra-LATA, long-distance, and international levels. The telephone network has become a network of networks.

Local telephone companies are restricted to providing service within their LATAs (intra-LATA service), while long-distance companies provide inter-LATA service.

A local telephone office is called an exchange. This terminology has its roots in the early days of telephone service where the human operators at the local telephone office, in effect, exchanged lines by forming the physical connections. Today, local telephone companies are called local exchange carriers (LECs), and the long-distance companies are called interexchange carriers (IXCs). Telephone calls that are billed in minutes of use are called toll calls. Toll calls made within a LATA are carried over intra-LATA toll networks.

Broadcast television signals are carried over coaxial cable into most homes in the United States. The provision of cable TV also involves networks—in this case, networks formed by the coaxial cable that is configured in the form of a tree architecture with trunks and branches formed by the cable as it makes its way to neighborhoods, ultimately terminating in homes at the cable box that connects to the home TV receiver. The cable TV network, unlike telephone and data networks, is one-way and is not switched. This could change in the future with new technological developments, and cable networks of the future might be switched, two-way networks, integrating together the delivery of voice, data, and video.

Public and private networks

Virtual networks are used by large businesses as a substitute for creating and maintaining their own private, physical networks. Decades ago, businesses would lease private line circuits from the established common carrier to create their own private networks for in-house use by their employees, usually connecting together various business locations. The circuits could be voice-grade private lines (VGPL), high-speed data lines, or, if the traffic were sufficient, a dedicated digital line capable of carrying 24 voice-grade circuits. A virtual network is created as a subnetwork of the public network, which means that public and private communication is all being transported over the same physical network, though separated in virtual space.

The telephone network, as mentioned earlier, is circuit (or virtual circuit) switched. Since the telephone network is available to everyone, it has become known as the public switched telephone network (PSTN). Here, the term "switched" is used to contrast the PSTN with networks created from private lines that are dedicated and then not switched. Today, as stated in the previous paragraph, private networks are switched and share space with public networks on the same physical facilities.

A systems perspective—the telephone network

The most visible component of telephone service, or telephony, is the telephone instrument itself. The telephone instruments in the home or business are all connected by wires that are referred to as intrapremises wiring. The telephone instruments themselves, along with such other devices as facsimile machines and telephone answering machines, are referred to as customer premises equipment (CPE). As shown in Figure 2.4, all the intrapremises wiring is brought together within the customer's premises at some central point. This central point then

Figure 2.4 A telephone call involves much switching and transmission equipment "beyond the dial." Each telephone is connected to a central office by a pair of wires, called the local loop. The first stage of switching occurs here at the serving central office. Calls to another office within the local access and transport area (LATA) are carried over interoffice trunks. Calls outside the LATA are handled by interexchange carriers (IXCs) over their own transmission and switching facilities. The point of presence (POP) is the place where the IXC connects to the facilities of the local exchange carrier (LEC).

connects with a single pair of wires, called the local loop, which connects the premises with a distant central location called the central office. The intrapremises wiring and customer premises equipment is protected from dangerous voltages, such as caused by lightning, by a fuse called the protector block, which is located at the point of connection of the intrapremises wiring to the local loop. The local loop is so called because there are two wires, which during the early days of telephony were visualized as forming a loop connection.

The wire pairs of many local loops from many premises are combined together in a thick cable. The cable might be fastened to telephone poles or buried underground. All the cables from a local area terminate at the central office that serves all the premises within its jurisdiction.

Equipment in the central office supplies many useful functions. The central-office equipment supplies a dial tone to signal the telephone user that the equipment is ready for dialing to commence. The equipment receives the dialed number, interprets it, and begins the steps necessary to make the connection over which the telephone conversation will occur.

If the called party is served by the same central office, the connection is made immediately and most simply. If the called party is located within the same immediate geographical area but is served by another central office, connection is first made over communication circuits called trunks between the central offices and then the final connection is made at the central office serving the called party.

If the called party is located far away in a distant part of the state or in another state, then the equipment in the calling party's central office will make a connection to the long-distance network. Equipment and facilities in the long-distance network will create a connected circuit to the distant central office serving the called party, where the final connection will occur. All this typically happens in a few seconds.

The place where the IXC connects to the LEC is referred to as the point of presence (POP). Some large users may connect directly to an interexchange carrier, thereby avoiding the local exchange carrier; this is called bypass. Bypass is supplied by competitive local exchange carriers (CLECs). The transmission facility over which telephone service is provided to a single user is called a circuit. A number of circuits are combined together for transmission between facilities over trunks, sometimes called trunk circuits.

During the early days of telephony (and today, too), a flat rate was charged that allowed the calling party to make unlimited calls within a local area. Calls outside that area were billed on the basis of the distance and the length of the call. These long-distance calls became known as toll calls, and the long-distance network was called the toll network. The telephone user subscribed to telephone service on a monthly basis, just as today, and became known as the telephone subscriber. The nation's toll network used to be operated solely by the Long Lines Division of the American Telephone and Telegraph Company (AT&T). Today, a number of firms provide long-distance service, either on their own facilities or by reselling service provided over another company's facilities.

The nation's long-distance network has become the synthesis of a number of separately operated facilities. Today's long-distance networks—and local networks, too—are controlled by computers and offer so many sophisticated features that they are often called intelligent networks.

A systems perspective—the intelligent network

A network is able to offer intelligent features because the signals that control it are carried over a separate dedicated network, called a common channel signaling network, as shown in Figure 2.5. The common channel signaling network is packet switched and carries the signals that control the network carrying customer traffic.

As shown in Figure 2.6, the signaling network communicates between signal processors that control the switching machines that perform the switching of the circuits and signals in the transport network carrying customer traffic. A switching machine is located at a service switching point (SSP), and a signal processor is located at a signal transfer point (STP). The local network (or LATA) connects to a signaling link at a signaling point of interface and to traffic trunks at a POP. A database of information about customers and their telephone numbers is located at a service control point (SCP). This database is accessed to determine information to enable calls to be directed to the appropriate destination, such as the translation of a toll-free number to the appropriate service number.

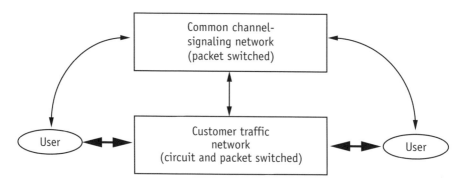

Figure 2.5   Intelligent networks offer features that are made possible through the flexibility of a separate common channel-signaling network that controls the switching and allocation of customer traffic.

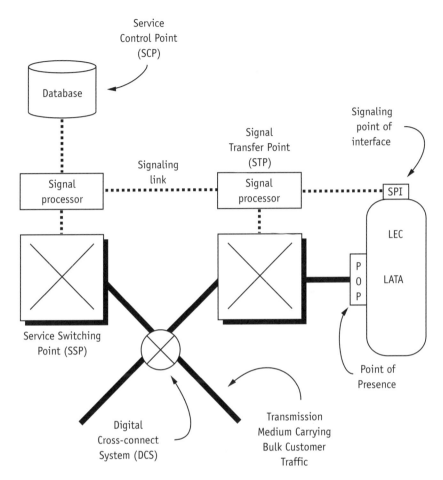

Figure 2.6 The signaling network communicates between signal processors that control the switching machines that perform the switching of the circuits and signals in the transport network carrying customer traffic. A switching machine is located at an SSP, and a signal processor is located at an STP.

Digital cross-connects

In the early days of telephony, trunks sometimes had to be interconnected. These interconnections were made with cables that would cross-connect one physical trunk to another. With modern digital trunks carrying thousands of voice circuits, such physical cross-connection was not efficient, and electronic cross-connection was invented to handle these high-capacity digital trunks. The earliest implementation in the early 1980s was called a digital access and cross-connect system (DACS). Today, it is called a digital cross-connect system (DCS).

A transmission medium carrying customer traffic can be cross-connected to another medium, usually at a bulk level of traffic, by a digital cross-connect system. Digital cross-connects obtain their instructions from their own dedicated packet-switched network (shown in Figure 2.7) under the direction of a facility monitor and control system (FMCS).

A DCS primarily is a way to perform electronic cross-connection, or redirection, of the signals from one physical trunk to another. Today, such trunks would usually be optical fibers carrying digital signals—what is known as a synchronous optical network (SONET). A DCS can reduce the number of physical trunks—an application called consolidation—by combining channels from a number of physical trunks onto a single trunk. Channels can be rearranged on a trunk—an application known as grooming. A number of channels can be combined (or multiplexed) onto a higher speed trunk.

Redundancy

Large metropolitan areas and major regions of the United States are served by networks configured in the form of large rings, or loops, so that a break in one section of a ring can be bypassed by going around another direction of the ring. In this way, service can be quickly restored in case of a break.

The maintenance of a high degree of service quality is essential given the critical importance of telecommunication in today's world. Service outages must be avoided, and should they occur, service must be reconnected in the shortest possible time. Most parts of the United States are served by loops so that if a break occurs in the transmission medium at one point in the ring, service can be maintained by an alternate path that avoids the break. This means that circuits have to be cross-connected from one transmission facility to another in a dynamic manner, and this is accomplished by the digital cross-connect systems situated throughout the network.

Figure 2.7 Transmission links carrying customer traffic are shown as solid lines; thick dashed lines represent common channel-signaling links to instruct switching machines at SSPs; and thin dashed lines represent signaling between DCSs.

Perhaps the most elaborate example of a modern telephone network is AT&T's long-distance network—a network that in 1997 handled over 200 million calls in a normal day and consisted of about 41,000 route miles of optical fiber. AT&T's network will be exemplified in the following description and is the basis for the diagram of Figure 2.7. AT&T's terminology is a little different, however. A digital cross-connect (DCS) is called a digital access and cross-connect system (DACS), a service switching point is called an action control point (ACP), and a service control point is called a network control point (NCP). The common channel signaling (CCS) network in AT&T's long-distance network is formed from the STPs and NCPs. The signaling links in the CCS network are from 1 to 16 data lines, each carrying two-way packet-switched data at 56 kbps. To achieve redundancy, the STPs operate in pairs with each STP in the pair being capable of doing all the work of the other in case of an emergency.

Customers wanting local access to AT&T's long-distance network do so at the LATA at a POP for traffic trunks connected to an appropriate action control point. To achieve redundancy, AT&T connections to each LATA are made at at least two POPs and associated ACPs. This way, if trouble occurs at one POP/ACP interface, access can be made at another POP/ACP interface. We already discussed the use of digital cross-connect systems to reconfigure and reroute trunks in case of a break anywhere in a loop. Clearly, DACSs are essential to the survivability of AT&T's network, and communication with them must be maintained by an alternative path different than the usual route. Accordingly, in addition to sending commands to the DACSs over optical fiber, commands can be sent over a communication satellite link and received at the DACS by a very small aperture terminal (VSAT) located at each DACS. Service can be restored very quickly with this arrangement, with the first channel (at 45 Mbps carrying 672 voice circuits) being restored within one minute and additional channels every few seconds after that. DACSs are usually located where one ring meets another.

AT&T has over 100 digital switching machines located at the various ACPs in its network. Processors and transmission paths in each switching machine are duplicated as a means of further redundancy in case of failure. Some of the switching machines at ACPs operate in pairs so that in case of an emergency one can take over the traffic being carried by the other. The traffic is routed in a dynamic fashion, with the requirement that no more than one intermediate ACP can be involved in the ultimate routing of a telephone call.

Networks—an assessment

The telephone network is what makes telephone service ubiquitous and accessible around the world. Voice networks and data networks will increasingly collaborate together with one increasing the value and utility of the other.

The classic way of describing the telephone network has been to examine each of its elements—station apparatus, transmission, switching, and signaling—individually in detail. The following chapters will do exactly that. But, these individual elements must all work together in a seamless fashion as a system to create a network providing useful services to its users.

Selected bibliography

Brosemer, J. J., & D. J. Enright, "Virtual Networks: Past, Present and Future," *IEEE Communications Magazine*, March 1992, pp. 80–85.

Cummings, J. C., K. R. Hickey, and B. D. Kinney, "AT&T Network Architecture Evolution," *AT&T Technical Journal*, May/June 1987, pp. 2–12.

Noll, A. M., "Network Security and Reliability: Emergencies In Decentralized Networks," in *Private Networks Public Objectives*, (E. Noam and A. NiShuilleabbain, eds.), Amsterdam, The Netherlands: Elsevier Science B. V., 1996, pp. 343–356.

Sable, E. G., and H. W. Kettler, "Intelligent Network Directions," *AT&T Technical Journal*, Summer 1991, pp. 2–10.

3

Station Apparatus

Introduction

The most visible aspect of telephone service is the telephone instrument itself. It is the telephone that we use to call and speak to other people. However, as we shall see much later in this book, there is considerably more "beyond the dial" to the provision of telephone service. Since it is so visible, we shall begin our understanding of the workings of telephone service by examining the workings of the telephone instrument itself.

In this chapter, we shall review the story of the invention of the telephone and its various component parts. It is surprising that the basic workings of the telephone have changed so very little over its century of history. But perhaps that is because the basic principles of converting sound to electricity and then back again are still much the same.

The telephone is only one type of apparatus that can be stationed on a customer's premises. Telephone answering machines and facsimile units are other types of station apparatus. Telephones and other station apparatus are considered customer premises equipment (or CPE).

The challenge

To this day confusion clouds who really invented the telephone. We all learn early in school that Alexander Graham Bell invented the telephone, but we do not know the name Elisha Gray. Historians now realize that Bell and Gray, working independently, both discovered the telephone, and although Bell obtained the patent, Gray seems to have invented it first. But one thing that is certain is that Bell and his financial backers were first to realize the commercial importance of human speech as a form of communication over distance—telecommunication—and to then develop it.

Back then in the second half of the 19th century, the telegraph was the means of communication over distance using copper wires to carry the on-off pulses of electricity. Many inventors were attempting to discover how to place more than one telegraph signal on each wire, including Thomas Alva Edison, among many others. One path of investigation was to use vibrating reeds tuned to different frequencies to carry different telegraph signals, a technique we today know as frequency-division multiplexing and about which much more will be said in a later chapter in this book. The quest for combining many telegraph signals on a single wire led to the question of whether there was a maximum to the number of signals that could be combined. The answer to the question of the maximum capacity of a communication channel came with the discovery of a mathematical theory of information by Claude E. Shannon in 1948.

The idea of vibrating reeds for telegraph signals led some inventors to wonder whether the vibrations of a speech signal could somehow be transmitted over a wire. Alexander Graham Bell—educated as a speech scientist and a teacher of the deaf—was well aware of the vibratory nature of speech signals. It is not surprising that he was one of the inventors of the how to do this, but so too were others (in particular, Elisha Gray).

The early failed attempts at converting speech to "undulating" or "vibratory" electric currents attempted to use a number of reeds or other vibrating elements tuned to different frequencies at both the transmitting and receiving ends. It took years before it was discovered that a single vibrating element could respond to all the frequencies in speech. What is interesting about the resonator approach is that it anticipated inventions that would come much later as a way to reduce the bandwidth needed to send a speech signal—a class of devices called vocoders, for voice coders. The many vibrating reeds were, in effect, performing a spectrum analysis of the speech signal.

But enough of these interesting historical connections and insights, and now on to our explanation of how a telephone works—the story of "behind the dial."

Bell's invention

March 10, 1876 is a very important date in the history of telephony, for it was on that day that the first working model of a telephone was demonstrated by Alexander Graham Bell and his assistant, Thomas A. Watson. The patent application for the telephone had been filed a month earlier on February 14, 1876 by Bell. However, a disclosure for a telephone had also been filed that same day by another inventor, Elisha Gray. Although Bell's invention was upheld in a split decision by the U.S. Supreme Court, mystery and controversy continue to cloud the question whether Bell had learned of and copied some of the key features in Gray's disclosure.

The key feature in the telephone demonstrated by Bell was the use of a variable-resistance transmitter, which generated an electric current that varied enough to generate an ample acoustic signal at the receiver. The idea of a variable-resistance transmitter was a key feature in the caveat—or warning to other inventors—filed by Elisha Gray on that fateful day of February 14, 1876. Bell's patent filing also disclosed the idea of a variable-resistance transmitter, but the disclosure was in the form of a handwritten marginal notation, almost as if it were an afterthought or last-minute addition. We will never know whether Bell, possibly through his supporters, had somehow learned of Gray's idea and added the marginal notation later or whether all of this was simply pure coincidence. But history certainly credits Bell with the invention of the telephone, and there is no doubt that Bell had the vision to foresee the vast business and social value of universal telephone service.

The telephone as invented and demonstrated by Bell consisted of a microphone, called a transmitter, and a small loudspeaker-like device, called a receiver, connected together by a pair of wires and in series with a battery. The series battery was essential to the concept of a variable-resistance microphone, as we shall soon see. Figure 3.1 shows a model of Bell's first telephone.

Bell's initial idea for a telephone system used an electromagnetic transmitter that was in essence a small loudspeaker-like device hooked up in reverse. The problem was that such an electromagnetic transmitter could not generate enough power to produce much sound at the receiver. The solution to this problem was the variable-resistance transmitter, in which the sound wave did not directly generate power but, rather, controlled the power flowing in an electric circuit.

As shown in Figure 3.2, the variable-resistance transmitter, the receiver, and a battery are all connected in series. The battery causes a

Figure 3.1   Photo of a model of the first electric telephone's liquid transmitter (on the left) and tuned-reed receiver (on the right). (Courtesy of AT&T Bell Labs.)

Figure 3.2 The variable-resistance transmitter was key to the working of the early telephone. The sound wave did not generate the power necessary to generate sound at the receiver but, rather, controlled the electric power flowing in the circuit. An external battery supplied the electric power.

constant electric current to flow through the circuit. That current I, according to Ohm's law, is the electromotive force E of the battery divided by the total resistance R in the electric circuit, or $I = E/R$. The total resistance comes mostly from the resistance of the microphone, since the receiver is a coil of wire (which has very little resistance). When the telephone user speaks into the transmitter, the increases and decreases in the sound pressure cause the resistance of the transmitter to likewise increase and decrease. These variations in resistance cause the current to decrease and increase, and the resulting variations in current are quite large and more than sufficient to generate a large acoustic output from the receiver.

The real problem for Bell was how to produce a variable-resistance transmitter. His solution, used in his first demonstration on March 10th, was the liquid transmitter. Sound waves were focused on a diaphragm to which a small wire was attached. The wire was immersed in a small metal cup containing a mixture of water and acid (see Figure 3.3). The wire moved up and down in the liquid in response to the sound waves, and as it did so, the electric resistance between the wire and metal cup decreased

Funnel to
concentrate sound

Diaphragm

Wire

Metal cup filled with
acid-water liquid

Figure 3.3   Bell's liquid transmitter consisted of a metal cup filled with a mixture of water and acid. The sound wave caused a small wire to move up and down in the liquid. The electric resistance between the wire and the cup thus varied in response to the sound.

and increased. A battery supplied a constant voltage to the circuit, and it—along with the varying resistance of the transmitter—produced the varying current necessary to drive the receiver.

Bell's first receiver was a reed tuned to vibrate in response to the varying current. Later, the reed was replaced by a coil of wire wound around a permanent magnet, as shown in Figure 3.4. The varying current in the coil of wire created a varying electromagnetic field, which attracted a metal diaphragm. This electromagnetic receiver was invented in 1876 by Watson, and the same principle continues to be used in many of today's telephone receivers.

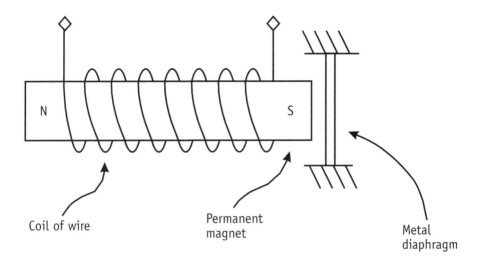

Figure 3.4 Watson's electromagnetic receiver was somewhat like a small loudspeaker. A coil of wire wound around a permanent magnet generated a varying magnetic field to create a varying attraction to a metal diaphragm.

Transmitters

During the early days of telephony, the key challenge was obtaining a signal of sufficient level so that the person at the other end of the connection could hear the speech. Because electronic amplification was not known then, the solution was to obtain as strong a signal as possible from the transmitter. Telephone instruments producing a larger output signal than others were judged superior, and thus the telephone company supplying those instruments had an advantage over its competitors.

The acid in Bell's liquid transmitter made it impractical for consumer use. Bell also investigated an electromagnetic transmitter in which a small magnet was attached to a diaphragm and the speech caused the magnet to move near a coil of wire, thereby inducing an electric signal. However, the output from this type of magneto transmitter was far too small.

In 1877, the "wizard of Menlo Park," Thomas Alva Edison, invented a variable-resistance transmitter utilizing a small disk of carbon in the form of compressed lampblack. The Edison transmitter produced a high output, and the non-Bell companies that used it had an advantage over the

Bell companies. However, the most important principle in developing a high-output transmitter was demonstrated in 1878 by David E. Hughes. Hughes demonstrated that one nail laid across two other nails, between which an electric circuit was connected as shown in Figure 3.5, would produce large variations in resistance when the loose nail was vibrated by an acoustic signal. He also demonstrated that such loose contacts with carbon would produce the same effect, and he revived the term "microphone" to describe a transmitter device based on the loose-contact principle.

A number of inventors investigated loose-contact microphones. Emile Berliner was one of them. Another was Francis Blake, who had joined the Bell company in 1878. Blake's loose-contact transmitter, invented that year, consisted of a platinum bead, held by a light spring, which vibrated against a hard carbon disk. The Blake transmitter then became the standard transmitter in Bell telephones. But more innovations in transmitters were soon to come.

On September 16, 1878, Henry Hummings received a British patent for a telephone transmitter using a cavity partially filled with finely divided carbon. The electric circuit was completed across the cavity. The American Bell Telephone Company purchased the rights to Hummings' invention, and in 1885 redesigned it to eliminate some of the packing

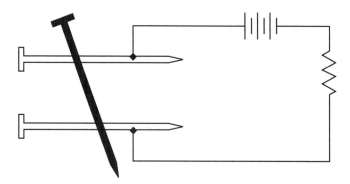

Figure 3.5　In 1878, David E. Hughes demonstrated that one nail laid loosely across two others would vibrate in response to an acoustic signal. That vibration would produce a varying resistance in the electric circuit formed across the two stationary nails.

problems that occurred with the carbon. Over time, the finely divided carbon simply became congealed, or packed, into a single mass, thereby defeating its use as a microphone.

In 1886, Thomas Alva Edison improved Hummings' invention through the use of roasted granules of anthracite coal in the cavity, as depicted in Figure 3.6. The many granules create many paths for the electric current, and hence the direct current (dc) supplied from the battery can be much higher than with previous designs. The diaphragm causes the carbon granules to vibrate against each other, thereby causing the resistance to vary.

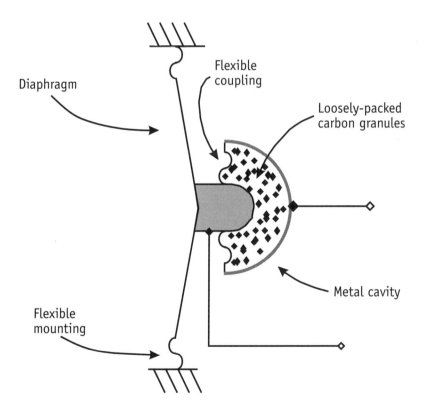

Figure 3.6   The carbon transmitter has a small metal cavity partially filled with carbon granules. The many granules are all in contact with each other and create many electric paths across the cavity. Since there are many paths, the current across the contacts can be higher.

The carbon-granule transmitter was a major innovation in telephony and is still encountered today, though improved in characteristics and manufacturability (see Figure 3.7). The carbon transmitter, or microphone, does not respond well to small acoustic signals and thus is good at discriminating against acoustic background noise. Furthermore, if the speech signal is very strong, the carbon microphone overloads somewhat, thereby protecting the telephone connection from excessive signal levels. These characteristics plus its low cost have accounted for the long life of this technology. The carbon granules can still stick, or pack, together, particularly in wall-mounted handsets, although a simple solution is to slam down the handset to loosen the granules.

Figure 3.7   Photograph of a modern carbon-granule transmitter showing the cavity and the carbon granules. (Courtesy of AT&T Bell Labs.)

Early ringers and dialers

Bell quickly realized that the person making a telephone call needed a way to signal the other person to pick up their telephone. People tapped their transmitter, but the signal at the other end was too weak to be heard. Bells and buzzers of various kinds were tried, but were not loud enough. Thomas Watson discovered the ultimate solution: a ringer consisting of a polarized call bell, shown in Figure 3.8. A permanent magnet supplied the energy to strike the bell hard enough to produce a loud sound. A pivoting metal armature was made to move by a varying electric current generated by a magneto in the calling telephone and operated by a hand crank. A hammer was attached to the armature, and it moved to and fro, striking two bells quite loudly. The patent application for Watson's ringer was filed on August 1, 1878. Today's telephone ringers with bells are essentially the same as the one invented by Watson.

Early telephone ringers were connected in series with the line and hence caused a transmission loss since the voice signal had to flow through the ringer, too. In 1890, a Bell employee, John J. Carty, invented a ringer with a high impedance of 1,000 ohms. Impedance is the opposition to the flow of alternating current in an electric circuit. Since Carty's ringer had a high impedance, it could be connected in parallel across the line and for voice frequencies would not draw appreciable current from the telephone or from the line, as shown in Figure 3.9. The high-impedance ringer required a higher voltage to operate it; an ac signal of 75V rms (root mean square) was used then and ever since. A capacitor is connected in series with the two ringer coils to prevent dc from flowing through the ringer.

At first, telephones were installed as private-line systems connecting only two instruments, as shown in Figure 3.10. In early 1878, the first telephone exchange was formally opened. It allowed 21 customers to reach each other by means of a central switchboard. The manual switchboard was operated by young men, and later by more polite and courteous women. An early switchboard is shown in Figure 3.11.

The first automated switchboard, installed in 1892, was invented by Almon B. Strowger. Strowger's automated switchboard consisted of an electromechanical switch that was operated by pulses from the telephone

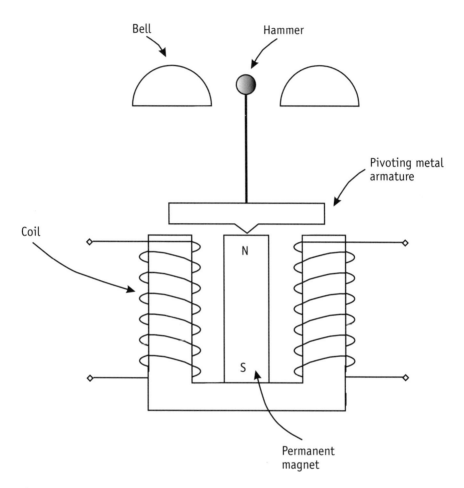

Figure 3.8   Thomas Watson invented the dual-polarized telephone ringer. A hammer is attached to a pivoting metal armature. The use of a permanent magnet increases the strength of the strikes of the hammer against the two bells. The two coils that carry the varying electric ringing signal are connected in series.

instrument. These pulses were produced when the user rapidly pushed a button on the telephone.

Strowger and his associates formed their own company, the Automatic Electric Company, to manufacture and sell automatic telephone systems. These systems were purchased and used by the non-Bell, or independent, telephone companies. It was not until 1919 that the first

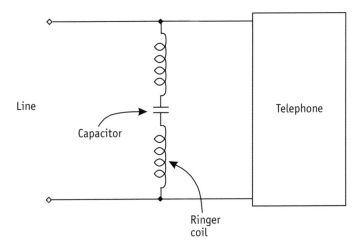

Figure 3.9 The telephone ringer is connected across the telephone line and in parallel with the rest of the telephone instrument. A capacitor is inserted in series with the two coils to block direct current from flowing through them.

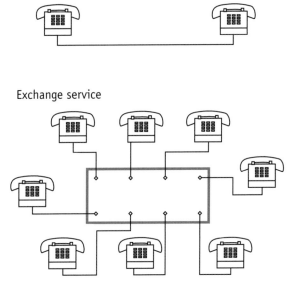

Figure 3.10 Private-line service simply connected together two telephones and no other telephones could be reached. With exchange service, any other telephone served by the exchange could be reached. Human operators made the cross connections from one telephone line to another.

Figure 3.11 Photograph of operators at work at a pyramid switchboard in Richmond, Virginia in 1881. (Photo courtesy of AT&T Bell Labs.)

Bell exchange was equipped with Strowger's automatic switching system. In 1896, A. E. Keith, J. Erickson, and C. J. Erickson, all associates of Strowger, invented a rotating fingerwheel for dialing the desired telephone number. Telephone dialing is explained later in this chapter.

The local loop

During the early days of telephony, a length of metallic wire connected the telephones, and the earth itself was used both as a source and as a sink for the flow of electrons that comprises electricity, as shown in Figure 3.12. Since nearly everyone, including the power companies, used the ground for electric circuits, considerable noise and interference resulted. The solution was devised by John J. Carty in 1881. He used a

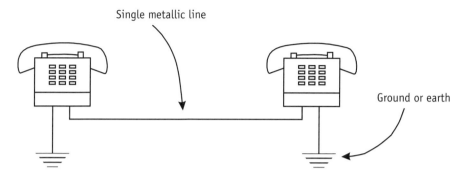

Single metallic line

Ground or earth

Figure 3.12   The use of the earth itself as a ground return for the flow of electrons that forms electricity created a source of considerable electric interference.

second metallic wire to complete the electric circuit, thereby avoiding the ground path, as shown in Figure 3.13. By slightly twisting together the two wires, noises from airborne electromagnetic interference was also reduced.

The two wires formed a loop between the telephone subscriber's premises and the local office of the telephone exchange. Thus, the wires were called the local loop. The term is still in use today to refer to the circuit between the subscriber's premises and the local office.

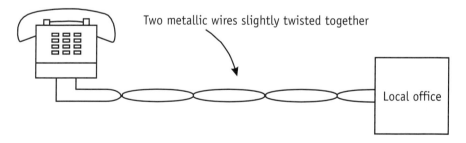

Two metallic wires slightly twisted together

Local office

Figure 3.13   The use of a second metallic wire to form an electric circuit was suggested in 1881 by John J. Carty. The circuit between the telephone at the customer's premises and the local office is called the local loop. The two wires are slightly twisted together to minimize the effects of airborne electromagnetic interference.

Power and ringing

A telephone is powered by direct current (dc). Early telephone instruments included their own internal battery to supply this direct current. In 1894, a common battery at the central exchange was used for the first time to power all the telephone instruments connected to the exchange. This common battery had a voltage of 48V. A common battery at 48V has been used ever since. The 48V are made negative with respect to ground to reduce corrosion of the copper wire loop.

When a telephone handset is on its cradle and the telephone is not in use, the telephone instrument does not draw any direct current from the central office. When service is desired, the user lifts the handset from its cradle, and a dc electric circuit is completed so that the telephone instrument draws dc over the local loop. The initiation of this flow of dc is sensed by equipment at the central office so that the party requesting service can be identified and served, as shown in Figure 3.14.

In early telephone instruments, the receiver was hung on a hook that activated a switch to control the flow of direct current to the instrument. This hook was called the switch-hook. The mechanism in today's telephone instrument that controls the flow of direct current is still called the switch-hook. The diagram of Figure 3.14 shows the connection of the switch-hook in a telephone instrument.

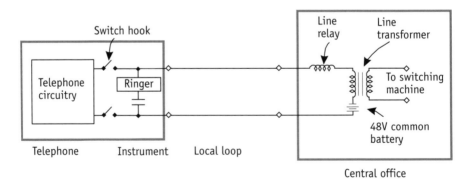

Figure 3.14   The circuitry in a telephone instrument that operates the instrument draws direct current from the local loop and is connected by the switch hook. The flow of direct current over the local loop is sensed by a line relay at the central office.

The telephone ringer is always connected across the telephone line. A capacitor in series with the ringer prevents direct current from flowing through it. The ringer has a high impedance, or opposition to the flow of alternating current, and thus little current from the speech signal flows through the ringer. The ringer voltage of 75V rms appears across the ringer and causes it to ring. The frequency of the ringer voltage is 20 Hz. The ringer voltage is on for 2 seconds and is then off for 4 seconds, as shown in Figure 3.15. When the called party answers the telephone by lifting the handset from its cradle, the switch-hook closes, direct current flows through the instrument, and central office equipment sensing that flow knows that the party has answered and stops the ringing signal.

Induction coil

The telephone first demonstrated by Bell was a one-way system. Telephone service was made two-way through the use of identical transmitter/receiver combinations in each instrument. The transmitter and receiver at each instrument were connected in series with each other and also with the battery that powered the transmitter, as shown in Figure 3.16. This arrangement had some serious problems.

The receiver typically had a much higher impedance than the transmitter. This greatly reduced the current flow in the circuit and, consequently, greatly reduced the received speech signal. The direct current

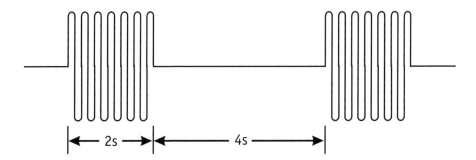

Figure 3.15 The ringing voltage consists of bursts of a pure tone, or sine wave, at a frequency of 20 Hz and with an rms electromotive force of 75V. The bursts are on for 2 seconds and off for 4 seconds.

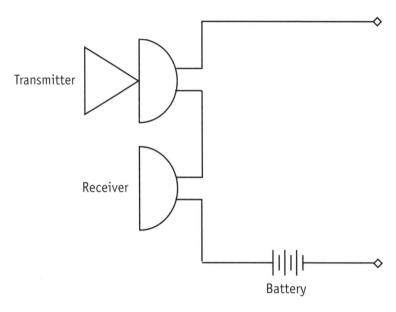

Figure 3.16   The transmitter, receiver, and battery were all connected together in series in early telephone instruments. Since the impedance of the receiver was more than the transmitter, the current in the circuit was reduced.

from the battery flowed through the receiver and could oppose the magnetic flux of the permanent magnet in the receiver, thereby reducing the efficiency of the receiver. The transmitter typically had an impedance of about 5 ohms, which did not match the impedance of the line (a few hundred ohms). This impedance imbalance reduced the energy transfer from the transmitter to the line. The solution to all these problems was the use of a transformer, initially called a repeating coil and later an induction coil, which was patented for this use in 1878 by both Berliner and by Edison.

An indicated in Figure 3.17, the transmitter and battery were connected in series with the primary of the induction coil. The secondary of the induction coil, the receiver, and the line were connected in series. An induction coil responds only to alternating current, and hence the direct current from the battery affected only the transmitter. The number of turns in the secondary winding of the induction coil was about 5 to 10 times the number of turns in the primary winding. This is called a step-up

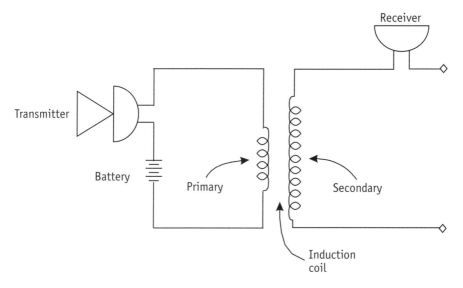

Figure 3.17 The induction coil, or transformer, was used to separate electrically the transmitter and receiver in a telephone instrument. The number of turns in the secondary was greater than in the primary so that the transmitter voltage was stepped-up before being sent down the line.

transformer, and it gives an increase in the voltage sent down the line from the transmitter. Furthermore, this type of transformer steps up the low impedance of the transmitter to match better the higher impedance of the line. The induction coil created a large improvement in the quality of early telephones. But an old problem remained—sidetone.

In early telephone instruments, the speech signal created by the transmitter passed through the receiver. Thus, a loud signal of one's own speech issued from the receiver, and this was particularly annoying when the receiver and transmitter were combined together into a single handset. The effect of hearing one's own speech while talking is called sidetone. As shown in Figure 3.18, we normally hear our own speech while talking through air conduction from the mouth to the ear and also through bone conduction. The electrical path from the telephone transmitter to the receiver creates a third path. A certain amount of sidetone is normal, and we use sidetone to adjust our speaking level. However, the amount of sidetone coming from an early telephone instrument was very loud, and users either lowered their speech or moved the telephone handset away.

Figure 3.18 While speaking on the telephone, we hear our own speech through air conduction from mouth to ear, through bone conduction, and also through an electrical path from the telephone transmitter to the receiver. The hearing of one's own speech while talking is called sidetone.

Either way, the transmitter signal was reduced, and the party at the other end received a much smaller signal; this was not an acceptable situation.

The solution was patented in 1918 by G. A. Campbell, an AT&T research engineer. Campbell showed that an induction coil could be connected in such a way as to balance or cancel the electrical sidetone. The actual antisidetone circuitry was not used commercially however until 1930 because of the need for exhaustive testing as well as the additional cost of the circuitry.

Antisidetone circuit

A center-tapped transformer is the basis of the antisidetone circuit, as shown in Figure 3.19. An electrical circuit, called a balance network, is needed to match or duplicate the impedance of the telephone line as seen by the telephone instrument. The receiver is connected to the secondary

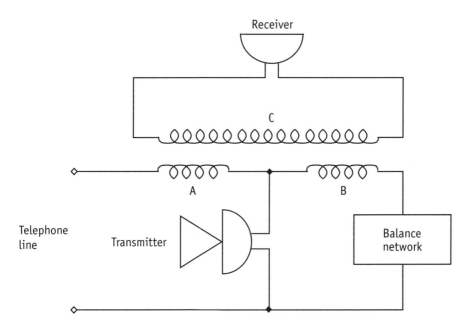

Figure 3.19 A center-tapped transformer is used in the antisidetone circuitry invented to reduce electrical sidetone to an acceptable level. The receiver is connected to the secondary winding of the transformer, C. The balance network attempts to duplicate the impedance of the telephone line as seen from the telephone instrument.

winding of the transformer. The transmitter is connected at the center tap of the primary winding. Assuming that the balance network perfectly matches the impedance of the line, the electric current generated by the transmitter will divide equally at the center tap and exactly half will flow in opposite directions through the primary windings. The induced voltages at the secondary winding have opposite polarities and cancel. Thus, no speech signal generated at the transmitter appears at the receiver. A signal coming from the line flows in the same direction through the primary windings in the transformer and induces a voltage at the secondary to activate the receiver.

Local loops vary in their impedances depending upon their length and other characteristics. Thus, no fixed-balance network can ever perfectly match exactly the impedance of the local loop. Therefore, an imbalance in the current flowing through the primary windings occurs, and some speech leaks to the receiver and is heard. As long as this leakage is not too much, it is acceptable and actually is desirable because some sidetone gives the telephone instrument a sense of being on. Many people blow into the transmitter to determine that the telephone is working. What they hear as a result is electrical sidetone.

Modern telephone instruments have circuitry in the balance network to compensate for different line lengths, and in this way, improved sidetone balance is obtained. However, some sidetone is desirable, and perfect balance, which could probably be achieved with today's electronic circuitry, is not desirable by reason of human-factor considerations.

The antisidetone circuit of present telephones is a little different from early telephones, as can be seen from Figure 3.20(a). The transmitter and center-tapped primary are the same. However, the receiver is connected with the balance network, and the secondary winding is depicted schematically, as shown in Figure 3.20(b), in such a way that the conventional transformer relationship is not readily apparent.

The local loop is a two-wire circuit. The transmitter and receiver in the telephone instrument form a four-wire circuit, as shown in Figure 3.21. The induction coil in the antisidetone circuitry in the telephone instrument performs this two-wire-to-four-wire conversion. Later in this book, we shall see how similar two-wire-to-four-wire conversion is needed for long-distance telecommunication. This conversion is performed by a transformer quite similar to the induction coil used in the antisidetone circuitry, except that it is called a hybrid coil. The induction coil in the telephone instrument is sometimes also known as a hybrid coil.

Dialers

The purpose of dialing is to specify the telephone number of the called party to the switching equipment at the central office. That number can be specified either by dial pulses or by touch tones. Dial pulses are produced by the rotary dialer.

The rotary dialer creates interruptions in the flow of direct current through the telephone instrument, and these interruptions, or pulses, are sensed and counted by the switching equipment at the central office. The rotary dialer is a mechanical device, as shown in Figure 3.22. As it is

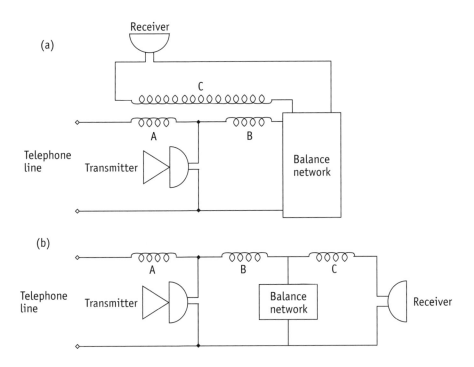

Figure 3.20 (a) In today's telephone instruments, the balance network and the receiver are connected together; (b) the actual schematic is usually drawn in such a way that the transformer relationship of the three windings is not readily apparent.

turned, a spring is wound. After it is released, this spring causes the dial to turn in the opposite direction, and a small governor controls its rate of rotation. While the dial is rotating during its return, it interrupts the flow of line current, thereby creating dial pulses at a rate in the range of 9 to 11 pulses per second. The line current is interrupted for about 60 percent of the time interval for each dial-pulse cycle. If the local loop is short, as is usually the case for the telephone lines to a private branch exchange (PBX), the dial-pulse rate can be increased to 20 pulses per second.

Although there are still some older rotary-dial telephones in use, most newer telephones use touch tones with push-button dialing. Touch-tone dialing is much faster than dial-pulse dialing. The push-button assembly is called a dial pad.

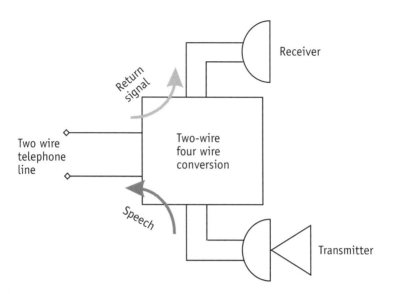

Figure 3.21 The two-wire local loop must be interfaced to the four-wire circuit formed by the transmitter and receiver in the telephone instrument. This two-wire-to-four-wire conversion is performed by the induction coil in the antisidetone circuitry.

Figure 3.22 The familiar rotary dial generates dial pulses during its return. These dial pulses occur at a rate ranging from 9 to 11 per second. A dial pulse is an interruption in the flow of direct current and lasts 60 percent of each pulse cycle. There is 0.7 second between dialed digits.

Each touch-tone digit is a unique combination of two single-frequency tones, as shown in Figure 3.23. The frequencies are arranged in a matrix. As the button is pushed for a specific digit, the appropriate combination of two tones is generated, corresponding to the intersection of the vertical and horizontal axes. The frequencies corresponding to the horizontal axis are called the low band and are 697; 770; 852; and 941 Hz. The frequencies corresponding to the vertical axis are called the high band and are 1,209; 1,336; 1,477; and 1,633 Hz. For example, if the digit 8 were pushed, two tones would be generated simultaneously and sent over the telephone line: one at 852 Hz and one at 1,336 Hz. The received tones are filtered and detected at the central office to determine the dialed digit.

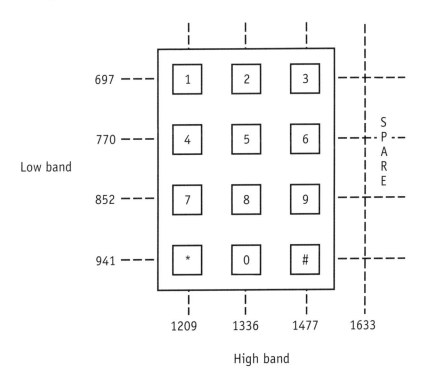

Figure 3.23 The dial pad used for touch-tone dialing. When a button is pushed, two tones at the frequencies corresponding to the intersection of the vertical and horizontal axes are produced. Equipment at the central office senses the frequencies of the tones and determines the dialed digit. A fourth vertical column at a frequency of 1,633 Hz is available but is not yet used.

Touch-tone dialing was initiated in 1963, and some telephone companies continue to charge extra for the service, even though touch-tone equipment is standard at the central office. Thus, a consumer might acquire a touch-tone telephone, but it might not work if the consumer has not ordered touch-tone service from the local telephone company. Some newer telephones have a dial pad with push buttons, and a switch selects whether dial pulses or touch tones are created when a button is pushed. These types of telephones work on any line, as do old-fashioned rotary-dial phones.

Circuit diagram

The major elements of a telephone instrument are the ringer, the switch-hook, the dialer, the transmitter, the receiver, and the antisidetone circuit (or hybrid). The overall relationship of these elements is shown in the block diagram of Figure 3.24.

A basic telephone instrument contains very little circuitry, compared to the complexity of today's other electronic devices. The circuit diagram for a telephone instrument is shown in Figure 3.25. The ringer is depicted

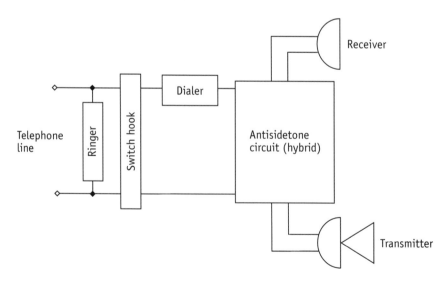

Figure 3.24   The major elements in a telephone instrument.

Figure 3.25 Circuit diagram of the GTE Starlite® telephone (type 182).

as a coil of wire, and the two lines next to the coil indicate that it has an iron core. The capacitor C_0 in series with the coil prevents direct current from flowing through the coil. The switch-hook is connected in parallel with the telephone line and is shown as two switches, SH1 and SH2. The switches are open when the handset is on-hook and are closed when the handset is off-hook.

A dial-pulse rotary dialer is shown as a switch, SD1. As the dial is released after being turned, the switch SD1 alternately opens and closes, creating interruptions, or pulses, in the flow of direct current. A capacitor, C_1, is connected across the contacts of the dialer to suppress the creation of sparks, which could otherwise interfere with radio and television reception. This capacitor also prevents the ringer from being weakly activated by the dial pulses during dialing, an effect called "tinkling." For this reason, the capacitor along with the resistor R_1 is sometimes called an "antitinkling filter."

The three windings of the center-tapped antisidetone transformer are shown in a fashion similar to Figure 3.20. The receiver is connected to the secondary of the transformer. Two switches are connected across the receiver. One of them, SH3, is closed when the handset is on-hook. Its purpose is to protect the receiver from any ringer current that might inadvertently leak into the receiver circuit and damage it. This switch opens when the telephone is off-hook and is in use. The second switch,

SD2, closes during dialing and prevents the dial pulses from being heard as loud clicks in the receiver.

A varistor (or variable resistor) consists of two diodes connected in parallel. It can be used in two different ways. If used one way, a varistor limits voltage excursions, and thus can protect the receiver from excessive signal voltages. For this purpose, the varistor V_3 is connected across the receiver and serves as a peak clipper.

If used another way, the resistance of a varistor varies inversely with the direct current flowing through it. The smaller the current, the higher the resistance. The direct current flowing down the telephone line varies with the length of the line. The longer the line, the less direct current is available to power the telephone instrument. If this effect were not compensated, the efficiency of the telephone transmitter would vary greatly with line length. The varistor V_1 is used to compensate for this effect, and it reduces the efficiency of the transmitter on short loops. The balance network for the antisidetone circuit is formed by C_2, C_3, R_3, and V_2.

The telephone connects to the line from two terminals, called the tip and the ring. This terminology comes from the early days of telephony when operators used a cord to connect one party's line to another party's line. The cord terminated in a plug, and the talking path was formed between the tip of the plug and a narrow ring around the plug. The terminology has remained.

Technological trends in telephone instruments

Telephone instruments continue to evolve in terms of reducing costs and offering new features to users. This section describes the technology behind some of these trends.

Figure 3.24 illustrated the major elements in a telephone instrument. Also included in a telephone are the housing of the instrument, the internal printed circuit board and components, and various cords and jacks (see Figure 3.26). Furthermore, modern features, such as a stored repertory of frequently dialed numbers, require some memory and processing in the form of a small microprocessor. A visual display is sometimes used to display telephone numbers and to make the enhanced features easier for the user to control.

Figure 3.26 Cutaway drawing of a Western Electric telephone instrument. (Courtesy of AT&T Bell Labs.)

Edison's carbon transmitter is still in use in some older telephones. The speech signal from the carbon transmitter can have a rough granular quality, but the rejection of background noise is quite good. Most telephones use miniature self-polarizing capacitive microphones, called electrets. The speech quality is much improved, and noise-reduction circuitry is used to reduce the deleterious effects of background noise. The receiver in most telephones is still based on the original electromagnetic principles first used by Watson. Some newer telephones use piezoelectric receivers that offer advantages in weight, cost, and efficiency.

The switch-hook is usually a simple mechanical switch. Electronic switches and miniature relays are sometimes used in newer telephones. The ringer and dc voltages encountered by the switch are fairly large and could damage many conventional integrated circuits. High-voltage transistors can be used to perform the switch-hook function, and newer technologies allow them to be dielectrically isolated from other elements on the same integrated circuit, a technique called hybrid integrated circuitry (HIC). The hybrid function is performed electronically in some newer telephones, thereby replacing the hybrid transformer.

The hammer-and-bell ringer invented by Watson is still used in many telephones. Newer telephones are using piezoelectric transducers and small loudspeakers. Electronic circuitry activated by the ringer signal creates the tones necessary for these transducers. The newer ringers are sometimes called alerters. They offer such advantages as small size, good efficiency, and nice sound characteristics. The circuitry required to generate the ringing signal is usually combined with the antisidetone network and the microprocessor on the same integrated circuit.

The old-fashioned rotary dial has faded away and has been replaced by push buttons on a keypad. The keypad usually controls circuitry to generate either tones for touch-tone dialing or dial pulses. Mechanical keypads have been replaced by membrane and rubber-dome technology.

The telephone instrument should be easy to use and should have a pleasing physical appearance. The human factors of telephone instruments continue to be important. Telephones should be "user friendly" in all respects.

Instrument variety

Telephone instruments come in all shapes and sizes with great variety. The simplest are black with a rotary dial and do nothing other than provide basic service. The complex are cordless electronic telephones with built-in memory for a large repertory of telephone numbers.

Cordless telephones consist of a base station connected to the telephone line and a cordless handset that uses radio transmission to and from the base station. Cordless telephones are great when on the patio sipping a martini by the pool. The elimination of the cord can be a great convenience while cooking in the kitchen, too. However, since radio transmission is used, anyone on the same frequency can listen to the conversation, thus making privacy an issue. Scrambling techniques are used on some newer cordless telephones to offer some protection against casual interlopers. The ringer is usually located in the base station, since if the ringer were in the handset it might ring while the user was about to make a call and could damage the user's hearing.

Some telephones offer hands-free operation. The speakerphone accomplishes this through the use of a separate unit containing both the microphone and the loudspeaker. The microphone also picks up room noise and reverberation, which gives the speech that "rain barrel" quality. Acoustic feedback can occur between the loudspeaker and the microphone, but the switching of electric loss into the feedback circuit is frequently used to prevent feedback. Acoustic feedback and reverberation arc eliminated in some newer speakerphones through the use of adaptive digital filters.

The simplest way to obtain hands-free operation is the use of a lightweight headset with the microphone located on a small boom. Such headsets are similar to those used by telephone operators and order takers. There is still a cord, however, so that physical movement is restricted. A number of newspaper reporters use such headsets while conducting interviews so that their fingers are free to type the interview into the word processor while the interview is being conducted.

A variety of "smart" features are available with many telephones. Last number redialing is a useful feature; the telephone remembers the number last dialed so that redialing a busy number is made much easier since only one key needs to be pushed. A repertory of frequently dialed numbers can be stored in the memory of some telephones and recalled simply by pushing a single key.

Station apparatus includes not only telephone instruments but anything that is connected to the telephone line. Facsimile machines for graphical communication, modems for computer communication, and telephone answering devices for recorded message taking (sometimes called voice mail) are some other station apparatus.

Station apparatus—an assessment

Some telephones resemble computers in all their buttons and sophisticated features and displays. I remember a repertory telephone that had a cathode-ray-tube display and could store thousands of telephone numbers. It was so complex that I doubt that I could have used it to make a simple call.

The telephone instrument is an essential telecommunication device that must work day to day and (particularly) in time of emergencies. A quality, dependable telephone is clearly essential.

The telephone instrument is basically a very simple device. New technology can be used to offer a great variety of sophisticated features, or it can be used to offer a simple, easy-to-use, quality product at a low price. All the most sophisticated features are useless if they are so complex that the average consumer cannot understand how to use them. I think of all the videocassette recorders that many consumers do not know how to program since they are too complex to fathom. Clearly, the marketing challenge is to develop in tomorrow's telephone instruments meaningful features that can be comprehended. The simplicity of Bell's invention contributed greatly to its success.

Selected bibliography

Hounsell, D. A., "Two Paths to the Telephone," *Scientific American*, January 1981, pp. 157–163.

4

Analog Transmission Systems

Introduction

The aspect of telephony that deals with the various media and technologies for conveying signals—be they speech, data, or video signals—from one place to another is called transmission.

A variety of different transmission media are used to transmit and convey signals. Sometimes these media carry only a single signal, while other times they carry many signals combined together, or multiplexed, through either frequency-division multiplexing (FDM) or time-division multiplexing (TDM). Frequency-division multiplexing is sometimes called analog transmission, and time-division multiplexing is sometimes referred to as digital transmission. Analog frequency-division multiplexing is the way of the past, and today most transmission is time-division multiplexing of digital signals.

Any transmission medium can carry signals that are combined by either frequency-division or time-division multiplexing. However, time-division multiplexing is usually associated with optical fiber. Analog transmission systems using frequency-division multiplexing are described in this chapter; and digital transmission systems utilizing optical fiber are described in the next chapter. However, please remember that although this book separates the transmission world into analog and digital, the boundaries are not that strongly defined. This chapter also treats the problems of echo that occur on all long-distance circuits, analog or digital.

Transmission media

The various transmission media are as follows:

- Open copper wire;
- Paired wire (commonly called twisted pair);
- Coaxial cable;
- Microwave radio (terrestrial and satellite paths) and;
- Optical fiber.

These media vary greatly in the number of individual telephone signals, or speech circuits, that they can carry.

Open wire and paired cable

Open wire consists of uninsulated, bare copper wires that are strung on poles. The two wires that form the pair needed for the transmission of electric signals are physically separated on glass insulators by a distance of about one foot (0.3m) to prevent short-circuits during high winds. Open wire is thick and therefore has a low loss, typically about 0.03 dB per mile. It was used during the early days of telephony until physical congestion became a serious problem. Open wire is still found, though very rarely, in rural areas.

A twisted pair is a pair of individually insulated copper wires twisted together with a full twist about every 2 to 6 inches. The insulation is usually plastic, but wood pulp has been used in the past. The diameter of the copper wires varies from 0.016 inch (26 gauge) to 0.036 inch (19 gauge). Many twisted pairs are combined together into a single cable, usually sheathed with plastic as shown in Figure 4.1, although older cables were often sheathed with lead. Anywhere from 6 to 2,700 twisted pairs are combined together into a single paired cable. The gauge of the wire varies with the number of twisted pairs in the cable; finer wire is usually used in larger capacity cables.

Paired cable can be strung on poles, buried underground, or installed in a conduit. A conduit consists either of long blocks of concrete with holes or of plastic pipe through which the cable passes. Conduit is buried underground and offers the advantage that cable can be easily replaced, without digging up city streets, simply by pulling out the old cable and pulling through the new cable.

The thinner wire used in paired cable has a higher loss than open wire. The heaviest gauge wire (19 gauge) has a loss of about 1.1 dB per mile at 1,000 Hz. The higher the loss, the smaller the signal that finally emerges

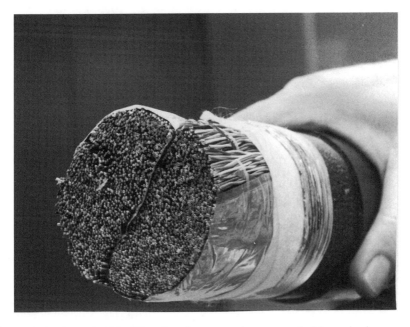

Figure 4.1 A total of 1,400 pairs of copper wire are carried in a single plastic-sheathed cable of about 3 inches in diameter. (Photo courtesy of AT&T Bell Labs.)

after travelling through the wire. Popular paired cables contain 2,700 pairs of 26-gauge wire, 1,800 pairs of 24-gauge wire, and 110 pairs of 22-gauge wire.

A problem with many wire pairs all running parallel to each other with close spacing in a cable is that the electric signal on one pair can leak to another pair. This effect is called crosstalk.

Paired cable is mostly used for the local loop, and also between local short-circuits exchange central offices. Baseband transmission is used on most local loops. In cases of high subscriber density and of very long loops, subscriber loop carrier (SLC) systems are available, using either frequency-division multiplexing via amplitude modulation (AM) or time-division multiplexing via digital techniques.

Analog FDM and digital TDM are used on paired cables in exchange trunk transmission when congestion necessitates an enhancement of baseband transmission. These multiplexing schemes are used on short-haul (15 to 200 miles) trunks. The analog systems are called N-carrier, and typically multiplex together 24 voice circuits. The digital systems are called T-carrier, and also multiplex together 24 voice circuits. Analog systems are very rarely in use today, and most interoffice trunks utilize the digital T-carrier technology. T-carrier is described in the next chapter on digital transmission systems.

Loading coils

A serious problem during the early days of telephony was how to receive a strong signal over long lines. These were the days before the dawn of the Electronic Age and the amplifier. The early solution was the use of reso-nance to boost the signal along the line by the deliberate introduction of series inductance in the form of coils of wire, called loading coils.

Loading coils were invented independently in 1899 by both Dr. Michael I. Pupin of Columbia University and by AT&T employee George A. Campbell. The final patent was awarded to Dr. Pupin based on a dis-closure only two weeks earlier than Campbell's. The theoretical analysis on which the inventions were based was performed by Oliver Heaviside in England in the late 1800s. The first loading coils were installed experi-mentally in 1899 and were quickly adopted for use on most long lines.

Even today, loading coils are used on long local loops, particularly in rural areas.

To understand why loading coils work, it is necessary to know that an electric transmission line can be modeled as a series of infinitesimally small elements, with each element consisting of a series inductance L, a series resistance R, a shunt (or parallel) capacitance C, and a shunt resistance S (see Figure 4.2). These various quantities are expressed in ohms, henrys, and farads on a per-unit-length basis—for example ohms per mile.

At frequencies encountered in a voice telephone call, the attenuation A (or loss) of such a line is given approximately by the following equation:

$$A \approx \frac{R}{2} \sqrt{\frac{C}{L}} + \frac{1}{2S} \sqrt{\frac{L}{C}}$$

The first term represents the effects of the series losses, and the second term represents the effects of the shunt losses. Usually, the series losses predominate. Thus, the introduction of additional series inductance will decrease the predominant first term, resulting in a decrease in the overall attenuation of the line. Clearly, if too much series inductance is added, then the second term could become predominant, thereby negating the desired effect of reducing the overall attenuation. The

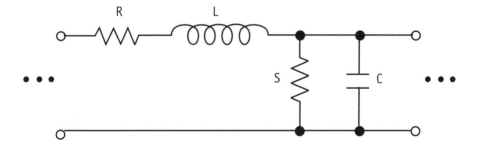

Figure 4.2 A transmission line is modeled as a series of very small sections, each section consisting of a series resistance and inductance and a parallel resistance and capacitance. These infinitesimal elements are usually expressed on a per-unit-length basis—for example, ohms per mile.

required series inductance is added as discrete inductors placed in series every 6,000 feet along the line. (See Figures 4.3 and 4.4.)

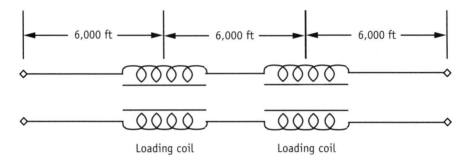

Figure 4.3   The series placement of loading coils reduces attenuation within the voice band. The loading coils are used on long loops and are spaced about every 6,000 feet.

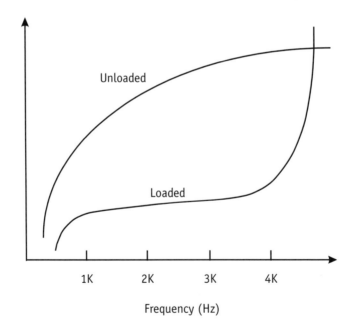

Figure 4.4   The effect of loading is to reduce attenuation in the voice band up to about 3,500 Hz. However, the attenuation increases rapidly above that frequency. The use of loading coils sharply reduces the higher frequencies outside the voice band that can be transmitted along the line.

An inductor is a coil of wire, and since these series inductors load the line to reduce attenuation, they are called loading coils. Their inductance is about 88 millihenrys. The inductance is chosen to ensure a passband from 300 to about 3,300 Hz. The type of loading most popularly used is specified as H-88. The H signifies a 6-kilofeet spacing, and the 88 signifies an inductance of 88 millihenrys.

Although a loading coil reduces attenuation in the voice band, attenuation outside this band is greatly increased. In essence, bandwidth is traded for reduced attenuation. The transition, or cutoff frequency f_c is given approximately by the following equation:

$$f_c \approx \frac{1}{\pi\sqrt{LDC}}$$

where L is the inductance of the loading coil, D is the spacing in miles between the loading coils along the line, and C is the shunt capacitance per mile.

Multiplexing

An early challenge to telecommunication inventors was how to transmit more than one signal over a single-wire circuit. The installation of wire circuits was very costly, and quite clearly great economies would result if more than one signal could be transmitted over each wire circuit. Among others, Thomas Alva Edison was attracted to solving this problem for telegraph signals; so, too, was Alexander Graham Bell in his investigations of the harmonic telegraph that ultimately led him to the invention of the telephone.

An early way for telephone signals to share wire circuits was called the phantom circuit. The additional signal was carried in a balanced way over existing wire circuit pairs through the use of transformers at each end of the circuit, as shown in Figure 4.5. The technique could be extended to more wire circuits so that more phantom signals could be carried.

Today, advanced techniques are used so that a single transmission medium can carry thousands of signals all combined together. The

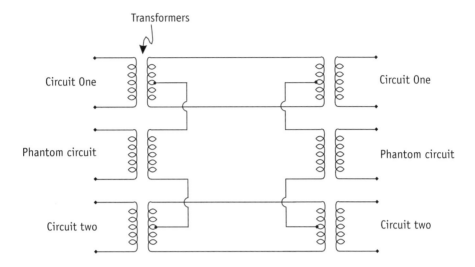

Figure 4.5   An additional phantom circuit is formed along two physical wire circuits through the use of transformers. The two wires of each physical circuit carry the phantom signal in a balanced way that has no net effect on signals carried over the physical circuits.

sharing of a transmission medium by many signals is called multiplexing, as illustrated in Figure 4.6. Two approaches to multiplexing are analog, or frequency-division multiplexing; and digital, or time-division

Figure 4.6   A multiplex unit combines together many voice (and possibly data) signals to share a transmission medium or channel. The multiplexing can be accomplished through either frequency-division or time-division multiplex units.

multiplexing. The actual equipment that performs the multiplexing is called a channel bank. Analog or A-type channel banks perform analog multiplexing; digital or D-type channel banks perform digital multiplexing.

Analog multiplexing

In analog multiplexing, a number of voice circuits are combined, with each voice circuit given its own unique space in the frequency spectrum. Each telephone speech signal occupies a frequency spectrum from about 200 to 3,400 Hz; such a signal in its most basic form is called a baseband signal. Each baseband speech signal needs to be shifted, or translated, in frequency to a new unique band of frequencies. This frequency shifting is accomplished by amplitude modulation using single-sideband, suppressed-carrier modulation. The actual multiplexing is accomplished as a multileveled process in which a small number of signals are multiplexed together to form a group, and then a number of groups are multiplexed together, and so forth.

A baseband telephone speech signal contains frequencies in the range from 200 to 3,400 Hz. A single voice circuit occupies a channel. Twelve channels are multiplexed together, each channel being 4,000 Hz wide, to create a group. A group covers the frequency range from 60 kHz to 108 kHz. Five groups are multiplexed together to create a supergroup occupying the frequency range from 312 kHz to 552 kHz.

The process can be continued (see Figure 4.7). Ten supergroups multiplexed together give a master group occupying the frequency range from 564 kHz to 3,084 kHz and containing a total of 600 voice channels. Six master groups multiplexed together give a jumbo group occupying the frequency range from 564 kHz to 17,548 kHz and containing 3,600 channels. Three jumbo groups multiplexed together give a jumbo group multiplex containing 10,800 channels.

Analog multiplexing and continual improvements in the technology enabled costly transmission systems to be shared to carry thousands of telephone signals, thereby making long-distance telephone service affordable to us all. But analog multiplexing suffered from noise, distortion, and other impairments, and was costly to maintain. The age of analog multiplexing is now over; digital multiplexing proved superior in

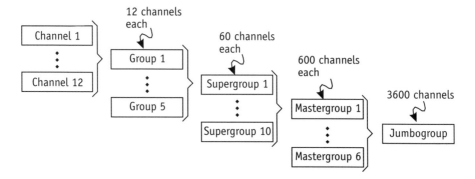

Figure 4.7   An analog multiplex unit, called an A-type channel bank, combines together voice signals starting with a basic unit of 12 voice circuits, called a group. Frequency shifting through single-sideband, suppressed-carrier amplitude modulation is used.

nearly all respects and is now quite widespread. In the late 1980s, AT&T replaced nearly all the analog multiplexing in use on its long-distance network with digital multiplexing systems; MCI followed in the early 1990s.

Coaxial cable

Analog multiplexing on coaxial cable has been used in the old Bell System since 1946 for long-distance telephone transmission. A number of one-way voice circuits are frequency-multiplexed together using single-sideband, suppressed-carrier amplitude modulation on a single coaxial pipe, or simply a coaxial. Two such coaxials make a two-way pair, with each coaxial carrying transmission in one direction. As the signal travels along the coaxial, it becomes weaker and weaker and must there be amplified before it becomes too weak. Amplification is a one-way affair, and thus a coaxial can only carry signals in one direction. A number of coaxials are placed together to form a coaxial cable for use in a transmission system. The photo of Figure 4.8 shows a coaxial cable used in the first-generation L-1 carrier system, installed around 1946.

The multiplexing system used with coaxial cable is called L-carrier. The various generations of the technology are indicated by a number

Figure 4.8 Cutaway view of the coaxial cable used in the L-1 system. Eight coaxial pipes formed the cable. (Photo courtesy of AT&T Bell Labs.)

suffixed after the L. A key factor in the L-carrier systems is the distance between the amplifiers, called repeaters, at which the signal is amplified and retransmitted down the next section of the cable. As the repeater and multiplexing technology improved, the distance between the repeaters could be decreased at a reasonable cost, and the overall capacity of the system could be increased.

The diameter of each coaxial is $\frac{3}{8}$ inch. Since the capacity of the system is large, one coaxial pair in the cable is kept as a spare to be used in case trouble is encountered with one of the pairs that are in service.

Table 4.1 shows the progression over time of the L-carrier system. The most recent system is L5E, which was first placed in service in 1978. The L5E system used integrated circuit technology with repeaters spaced every mile along the route. As with all L-carrier systems, the cable is buried underground. The channel capacity of a single coaxial in the L5E system is 13,200 channels (22 master groups). There are 11 coaxial pairs, 10 of which are in actual service. Thus, the overall capacity of L5E system is 132,000 two-way voice circuits. Coaxial cable is being replaced by optical fiber.

Table 4.1
The Generations of the L-Carrier Coaxial Cable System

	L1	L3	L4	L5	L5E
Service date	1946	1953	1967	1974	1978
Technology	Vacuum tube	Vacuum tube	Transistor	Integrated circuit	Integrated circuit
Repeater spacing (miles)	8	4	2	1	1
Capacity per coax					
in channels	600	1,860	3,600	10,800	13,200
in groups	Master group	3 master and 1 subgroup	Jumbo group	Jumbo group multiplex	22 master groups
Coax pairs	4	6	10	11	11
Working pairs	3	5	9	10	10
Route capacity (two-way voice circuits)	1,800	9,300	32,400	108,000	132,000

Microwave terrestrial radio

Basic principles

Radio transmission is used to carry telephone conversations across continents and oceans. Different frequency bands have been allocated for use by the common carriers for the transmission of telephone signals. A large number of voice circuits are multiplexed together in these radio systems.

The microwave radio bands are in the gigahertz (10^9) range of frequencies and are used in cross-country terrestrial and satellite routes. The extremely high frequency waves are transmitted between radio antennas. The radio waves are conducted to and from the antennas through metal pipes called waveguides. Two microwave-radio bands in current use are from 3.7 to 4.2 GHz, called the 4-GHz band, and from 5.925 to 6.425 GHz, called the 6-GHz band. The width of each of these bands is 500 MHz, and they are both used for microwave terrestrial radio. Each band is subdivided into a number of channels. The width of each channel in the 4-GHz band is 20 MHz, and the width of each channel in the 6-GHz band is 30 MHz.

Microwave radio beams follow a line-of-sight path. This means that radio towers for the antennas need to be located about every 26 miles along the route of the system to take into account the curvature of the earth, as shown in Figure 4.9. An antenna on each tower receives the radio signal and the received signal is shifted to a new channel, amplified, and retransmitted to the next tower (see Figure 4.10). The towers perform as repeater stations. For the 4-GHz and 6-GHz bands, rain does not usually have a significant effect on the propagation of the radio wave. However, rain does have a significant effect on the higher frequency 11-GHz (10.7 to 11.7 GHz) and 18-GHz (17.7 to 19.7 GHz) bands, which are also used for microwave transmission.

Microwave terrestrial radio used to be the backbone of the long-distance network, and is still a fairly important technology. Analog FDM was used extensively with microwave radio systems in the past. The increasing use of optical fiber has supplanted microwave radio to a considerable extent, but existing microwave routes have been converted to digital time-division multiplexing, thereby giving renewed life to an old

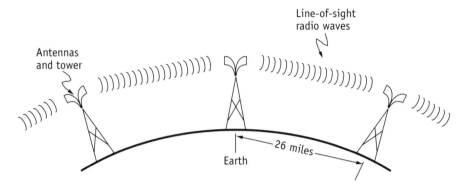

Figure 4.9   In terrestrial microwave radio systems, radio signals travel from tower to tower, each tower located a line-of-sight 26 miles from the next.

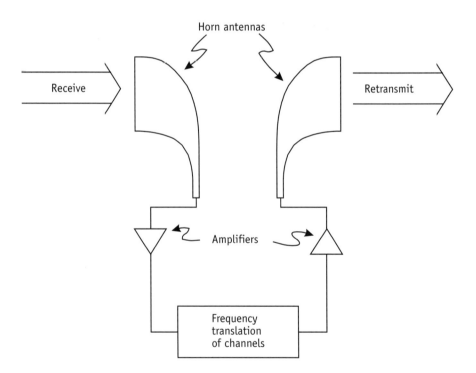

Figure 4.10   At each microwave tower, the received signal is changed in frequency to a different radio channel, amplified, and retransmitted to the next tower.

technology. We now review the history of microwave radio. It is an exciting story of how advances in technology continually led to ever increasing channel capacity and lower rates to the consumer.

TD radio system

The first cross-country microwave radio system in the AT&T long-distance network became available in 1950. It used the 4-GHz band and was called TD-2. The 500-MHz bandwidth was divided into 25 channels, each 20 MHz wide (see Table 4.2). Because of mutual interference, adjacent channels could not be used. Two channels were needed to create a two-way circuit with one channel in each direction, and thus the maximum number of two-way channels was only six. Five of these six two-way channels were used for service, and one was reserved as a spare.

Noise immunity was a problem, and hence frequency modulation with its greater immunity to noise and interference was used to modulate the radiofrequency carrier for transmission. However, frequency modulation trades noise immunity for bandwidth, and hence only 480 voice circuits could be transmitted over each radio channel. The overall capacity of the TD-2 system was five two-way radio channels with each channel carrying 480 voice circuits for a total system capacity of 2,400 two-way voice circuits. In 1953, the capacity of each radio channel was increased to 600 voice circuits for a total system capacity of 3,000 two-way voice circuits. This total capacity was soon doubled through an advance in technology.

Table 4.2

Two Microwave Bands Most Often Used for
Long-Distance Transmission Since They Are Least Affected by
Propagation and Noise Problems

	Frequency Range	Bandwidth	Channel Width	Number of Channels
4 GHz Band	3.7 to 4.2 GHz	500 MHz	20 MHz	25
6 GHz Band	5.925 to 6.425 GHz	500 MHz	30 MHz	16

Radio waves can be polarized so that they propagate as either vertical or horizontal waves. By polarizing one radio channel vertically and the adjacent radio channel horizontally, interference between adjacent channels was eliminated, and all 24 radio channels were utilized. This dual polarization was accomplished with the invention of a special horn antenna, shown in the photograph of Figure 4.11. The net effect was to double the total system capacity in 1959 to 6,000 two-way voice circuits. The truly exciting aspect of this development was that the same physical tower and route were used, simply be adding new antennas and other electronics.

In 1968, a solid-state transistorized microwave radio system operating in the 4-GHz band was introduced. This system, called TD-3, had

Figure 4.11   Photograph of a microwave relay tower used in AT&T's long-distance network. (Photo courtesy of AT&T Bell Labs.)

1,200 voice circuits per radio channel and a total system capacity of 12,000 two-way voice circuits. The final version of TD technology using analog FDM of the voice circuits had 1,800 voice circuits per radio channel. Of the total of 12 two-way radio channels, 11 were in service and 1 was for protection, giving a total capacity of 19,800 two-way voice circuits.

TH radio system

The horn antenna first used in the late 1950s was capable of handling radio signals in both the 4-GHz band and the 6-GHz band. Hence, a microwave radio system operating in the 6-GHz band could "piggyback" on the existing 4-GHz TD system. The 6-GHz system was called TH radio.

The first TH system became available in 1961 and utilized vacuum-tube technology. As shown in Figure 4.12, by utilizing dual polarization, all 16 radio channels could be used. This gave a total of eight two-way radio channels, six of which were in service. The TH system had a capacity of 1,800 voice circuits per radio channel and a total system capacity of 10,800 two-way voice circuits. The TH system was used with an already existing TD system for a total combined capacity on the route of 30,600 two-way voice circuits. In 1979, improved amplifiers and other technological advances enabled the capacity of the TH system to be increased to 2,400 voice circuits per radio channel. But this improvement was short-lived.

AR6A radio system

In 1981, AT&T introduced the AR6A system as a replacement for the TH system. The major innovation in the AR6A system was the use of a single-sideband, suppressed carrier microwave transmission of the radio signal with its inherent increase in bandwidth efficiency as compared with the previously used frequency-modulation transmission. This meant that with the AR6A system 6000, voice circuits could be placed in each radio channel.

The total of seven active two-way radio channels available in the 6-GHz band, with one two-way channel for protection, gave a total system capacity of 42,000 two-way voice circuits. Because AR6A was used in addition to the existing TD system, the total maximum route capacity

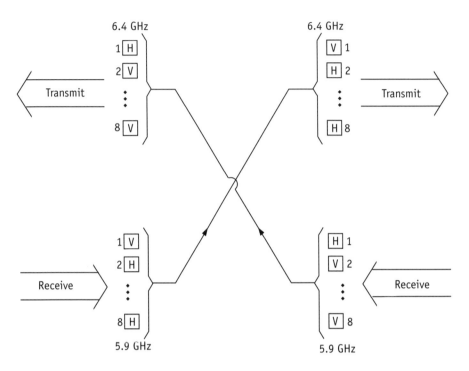

Figure 4.12   The use of vertical and horizontal polarization of the radio waves doubled the capacity of a microwave radio system by reducing interference between adjacent channels. In the TH system, the signal received in the 5.9-GHz portion of the band is retransmitted to the next tower in the 6.4-GHz portion of the band.

equaled the 42,000 circuits of AR6A plus the 19,800 circuits of TD for a total of 61,800 two-way voice circuits. However, the days of frequency-division multiplexing were soon to be over, and this advance would be supplanted by digital time-division multiplexing, covered in the next chapter.

Short-haul systems

The number of retransmissions, or hops, of the microwave radio signal distinguish short-haul microwave systems from long-haul systems. Short-haul systems typically have from 1 to 10 hops, and long-haul systems have from 10 to 50 hops. Microwave systems that cross the continent are long-haul systems.

Microwave radio is also used for short-haul communication utilizing the 6-GHz and 11-GHz bands. The 6-GHz system is called TM, and the 11-GHz system is called TL. Each radio channel in these systems carries 1,200 voice circuits utilizing analog frequency-division multiplexing. To prevent interference between the received and retransmitted signals, 12 channels in the 11-GHz band are used for a given hop, and the remaining 12 channels are used in adjacent hops. The net result is that six two-way channels are available for a total capacity of 7,200 two-way voice circuits for the 11-GHz TL system.

Communication satellites

Basic principles

Terrestrial microwave radio cannot go everywhere. It is very costly to construct a series of microwave towers to reach remote areas with few telephone subscribers. Also, terrestrial microwave radio cannot cross oceans because radio waves travel along a lone-of-sight path and floating microwave towers are not feasible. The solution is a single microwave "tower" placed high enough in the sky so that great distances can be covered in a single hop up to and down from the tower. Since neither skyhooks nor towers hundreds of miles high are feasible, some practical form of implementation is needed.

The practical solution is the use of an artificial earth satellite for communication purposes. This solution was first proposed by the renowned science fiction author Arthur C. Clarke in 1945 (in *Wireless World*). Scientists at Bell Labs and elsewhere became intrigued with Clarke's idea. By the end of the 1950s, a team of scientists at Bell Labs under the technical direction of John R. Pierce had constructed some prototype systems. The first communication satellite, Telstar I, was launched on July 10, 1962.

The early system proposed by Bell Labs would have consisted of a large number of satellites at the relatively low altitude of 3,000 miles. At this altitude, satellites orbit the earth very quickly, and hence a series of satellites would need to be tracked continuously across the sky as they passed overhead. Such a complex system had serious technical and economic problems. The more practical solution was a single satellite orbiting the equator at a height such that the orbit time of the satellite was exactly the same 24 hours as the daily rotation of the earth. In this way, the satellite would appear stationary with respect to the surface of the earth if the satellite were in an orbital plane with the earth's equator.

The physical principles that explain satellites are fairly simple. (See Figure 4.13.) Imagine a cannon located at the top of a tall tower and aimed parallel to the ground. If a shot is fired from the canon, the shot will travel a distance from the cannon the same time as gravity pulls it back to the ground. If more charge is added to the cannon, the shot will travel farther, but will ultimately fall back to the ground because of gravity. If a sufficient charge is placed in the cannon, the shot will have enough velocity so that the curvature of the earth's surface will be falling away at the same rate as the shot is being pulled toward the ground by gravity. The shot will

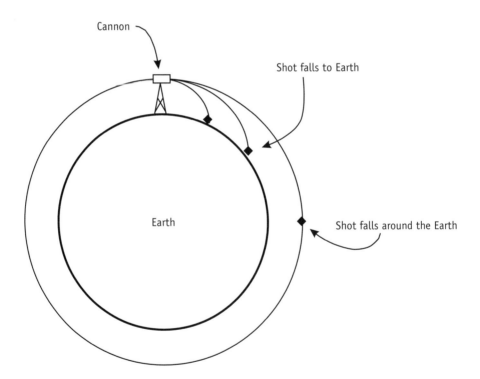

Figure 4.13 A communication satellite continuously falls back to earth but at the same rate as curvature of the earth falls away from it. The satellite thus orbits the earth.

then continuously fall around the earth. It will have become an artificial satellite of the earth.

The height of the orbit of a satellite and its rate of rotation about the earth are related. The higher the orbit, the slower the rate of rotation about the earth. There is a height at which the rate of rotation of the satellite is exactly the same 24 hours that it takes for the earth to complete one full revolution about its axis. If the orbit of the satellite is exactly above the earth's equator, is circular, and is in the same direction as the earth's rotation, the satellite will appear stationary with respect to the earth's surface. This type of orbit is called a geostationary orbit. The height of such a circular orbit is 22,300 miles above the earth's equator, or equivalently 26,300 miles from the earth's center (see Figure 4.14).

A communication satellite in a circular orbit 22,300 miles above the earth will be in synchrony with the 24-hour period of revolution of the earth. Such geosynchronous orbits are not necessarily in a plane with the earth's equator and hence are not necessarily geostationary. A satellite in an orbit inclined to the plane of the equator will appear to drift above the earth's surface and thus will need to be tracked by an antenna on the earth.

As stated earlier, the use of geostationary satellites for communication was first suggested by Arthur C. Clarke. The first geostationary satellite for intercontinental communication, Syncom II, was launched in July 1963. This satellite spun about its axis to achieve stability in orbit and was conceived in 1959 by Harold A. Rosen and designed by his team of engineers at Hughes Laboratories.

Electronic circuitry on the communications satellite receives the signals transmitted to it from the earth station. These signals are very weak and must be amplified by low-noise amplifiers (LNAs) carried on the satellite. The signals are then shifted to a new frequency band and are retransmitted back to earth. The frequency bands are divided into a number of radio channels, and a separate radio signal is sent in each channel, as with terrestrial microwave transmission. The amplification, frequency translation, and retransmission of the signal for each channel are preformed by circuitry called a transponder on the satellite. Each radio channel has its own transponder, and hence the satellite must have a number of transponders to cover the whole frequency band assigned to it.

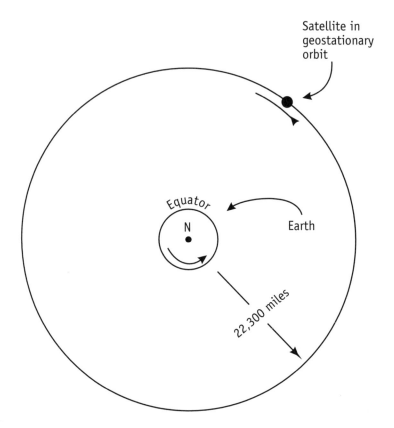

Satellite in
geostationary
orbit

Equator

N

Earth

22,300 miles

Figure 4.14 A communication satellite in a circular orbit 22,300 miles above the earth's surface takes 24 hours to complete one full orbit. If located in a plane above the earth's equator, it thus appears stationary with respect to the earth turning at the same rate beneath it. This is called a geostationary orbit.

Modern communication satellites operating in the most popular frequency bands typically carry 24 transponders.

Frequency bands

High-frequency microwave transmission is used to send signals up to communication satellites and then back to earth. The microwave band of frequencies used to transit from the earth to the satellite is called the uplink, and the band of frequencies used to transmit from the satellite back to the earth is called the downlink, as shown in Figure 4.15. The frequency bands must be different to eliminate interference between the uplink and the downlink.

The earliest frequency bands used for satellite transmission were the same 4-GHz and 6-GHz bands used for terrestrial microwave transmission. The 4-GHz band (3.7 to 4.2 GHz) is used for the downlink, and the 6-GHz band (5.9 to 6.4 GHz) is used for the uplink. The 4-GHz and 6-GHz bands used together for satellite communication are called

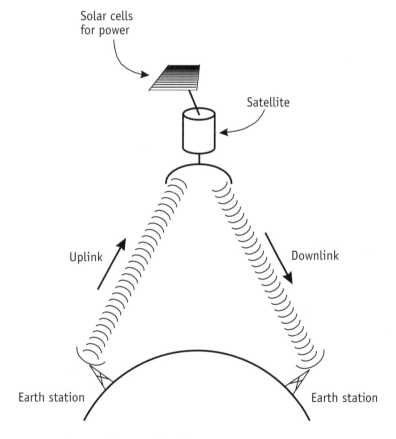

Figure 4.15   The signal transmitted from an earth station to a communication satellite is called the uplink, and the signal transmitted back to earth is called the downlink. Usually, a single antenna on the satellite with different feeds is used for both the uplink and the downlink.

the C band, or sometimes the 6/4-GHz band. Newer communication satellites use the Ku band operating at 12 GHz (11.7 to 12.2 GHz) for the downlink and 14 GHz (14.0 to 14.5 GHz) for the uplink. International agreements have authorized the use of a third band, the Ka band, at 20 GHz for the downlink and 30 GHz for the uplink.

Because only a few bands are available for satellite communication, the same bands must be shared by all communication satellites. Thus, to prevent interference, it is essential that the radio signal sent to one satellite not be received by an adjacent satellite operating in the same band, and similarly the earth station aimed at one satellite must not receive the signal from an adjacent satellite. This means that the transmitted radio beam and the aperture of the receiving antenna must both be very narrow. However, these widths depend on the frequency band used for transmission and are a factor in determining the permissible spacing between the satellites in geostationary orbit (see Figure 4.16). The

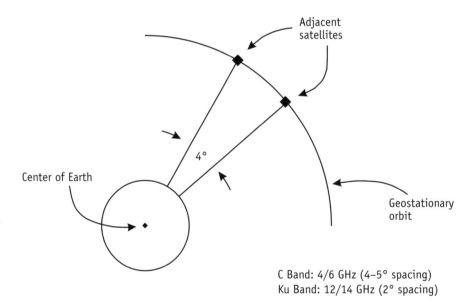

C Band: 4/6 GHz (4–5° spacing)
Ku Band: 12/14 GHz (2° spacing)

Figure 4.16 Communication satellites are spaced apart in geostationary orbit. The angle subtended with the center of the earth is about 5 deg for the 4/6-GHz C-band. The use of the higher frequency Ku band allows the spacing to be reduced to 2 deg, but propagation problems are much worse in this band.

permissible spacing in the C band has been between 4 to 5 deg, which translates into a distance about 2,000 miles between adjacent satellites. The high-frequency Ku and Ka bands allow the beam of the radio signal to be more narrow than the C band. With the Ku band, the distance between the satellites can be decreased to 2 deg, therefore allowing more satellites to serve the same geographic area. However, the higher frequency bands are more subject to absorption of the radio signals by rain.

The typical spectrum width of the radio channel served by one transponder in the C band is 36 MHz, with an additional 4 MHz for some guardband space thereby giving a total bandwidth per channel of 40 MHz. Because reception of a weak signal in the presence of noise is a prime consideration in satellite communication, frequency modulation with its inherent noise immunity is used for modulating the microwave radio signals. A single transponder can be used for one color broadcast television signal, 1,200 voice circuits, or digital data at a rate of 50 Mbps.

The total width of each half of the C band is 500 MHz. Horizontal and vertical polarization of the radio signal is used so that 24 channels are available for use. A communication satellite allocates these channels to create 12 two-way transponder pairs.

Delay

Geostationary communication satellites are at a considerable distance above the surface of the earth, and hence a fair amount of time is required for the radio signal to reach the satellite from the earth and to be retransmitted back. Radio waves travel at nearly the speed of light, which is 186,000 miles per second. A radio wave will therefore take about 120 milliseconds (ms) to travel the 22,300 miles that the satellite is above the earth. Since the earth station is not directly below the satellite, a little extra time needs to be added for this extra distance. Thus, a net time of 135 ms is reasonable for the radio wave to travel between the earth station and the satellite.

The signal received by the satellite is retransmitted back to earth. An additional 135 ms is required for the retransmitted signal to travel that distance. Thus, there is a total time of about 270 ms for a signal to be sent by satellite from one location on the earth's surface to another, assuming that a single satellite hop is used.

For a telephone conversation, one person speaks and the other person responds. Using a satellite, it would take 270 ms for the speech to travel from one person to another and an additional 270 ms for the response to travel back. The total round-trip time before the first speaker hears the beginning of the response thus is 540 ms, or about one half second. This amount of time is perceived as a delay in response, is quite noticeable to most people, and is quite bothersome to many. If one were using a computer terminal to a distant computer via a satellite link, this round-trip delay would add appreciably to the response time. This delay would be intolerable if one computer were communicating interactively with another at high speeds.

In some communication situations, one satellite hop might be connected back to back to another satellite hop, thereby doubling the round-trip delay. The delay in such multihop satellite links can bring interpersonal communication to a virtual halt unless each person learns to wait patiently for the other person's response and never interrupt the other person. A possible solution to multihop links would be to beam signals directly from one satellite to another, thereby eliminating the need for intermediate retransmission to the earth. The use of lasers would be particularly applicable to this type of intersatellite communication because there is no optical interference in the vacuum of space.

Access

The size and shape of the satellite's radio beam on the surface of the earth is called the footprint of the satellite. Clearly, for a satellite that is broadcasting signals to as large an audience as possible, such as in direct broadcast satellite (DBS) transmission of television signals to homes, the footprint should be large. If the satellite is sending signals to only a specific earth station, then the footprint should be small. In the first case, the radio beam would need to be broad; in the second case, the radio beam would need to be narrowly focused. The antenna design and available transmitting power on the satellite are important factors in determining footprints.

The satellite's antenna can be designed so that it can send narrowly focused beams to a number of specific, geographically dispersed locations. Furthermore, the narrowly focused beam can be switched from one location to another so that information can be transmitted to one location and then to another. If broad beams are used to reach multiple locations, then everyone receives the same information and security may be a problem. Narrowly focused beams, switched or unswitched, avoid this problem.

A number of separate earth stations may wish to beam information to the same satellite. One way this can be accomplished is for each multiple user to be assigned a specific range of frequencies within a transponder channel for each transmission. This type of access is called frequency-division multiple access (FDMA). Another way to accomplish multiple access is to assign specific times, or time slots, to each user during the time the information must be transmitted. This type of access is called time-division multiple access (TDMA), and is particularly well suited to the bursty nature of data communication.

In addition to the actual communication signals relayed by a satellite, telemetry signals are transmitted to and from the satellite. These signals report on the satellite's temperature, battery life, orbital position, and other parameters. If the position of the satellite needs to be corrected because of drift, signals from the earth will activate small thrusters on the satellite to reposition it.

Satellites are powered by solar energy from light rays from the sun. This light energy is converted into electric energy by panels of solar cells attached to the satellite. The efficiency of this solar conversion of energy determines the total electrical power available to the satellite for transmission of signals back to the earth, among other purposes. For a portion of each orbit, the satellite will be shielded from the sun by the earth. During this time, batteries previously charged from the solar panels are used to power the satellite.

Capacity

The communication satellite *Comstar I* operating in the 6-GHz C band was launched in 1976. It carried 24 transponders, and each transponder channel had a capacity of 2,400 voice circuits. Using horizontal and vertical polarization of the radio waves, the total capacity of the satellite was 14,400 two-way voice circuits. *Comstar IV* was launched in 1980. Its total capacity was increased to 21,600 two-way voice circuits through improvements in technology that enabled each transponder channel to carry 1,800 voice circuits.

Broadband frequency modulation was originally used for the radio signals transmitted to both *Comstar I* and *Comstar IV*. In 1982, the equipment at the earth stations was changed to use single-sideband amplitude modulation (SSB-AM) of the radio wave. SSB-AM is considerably more spectrum efficient than broadband frequency modulation, and hence the number of voice circuits handled by each transponder was increased to 7,800 for a total capacity of 93,600 two-way voice circuits. This significant increase in capacity was equivalent to that achieved through the similar use of SSB-AM in the AR6A terrestrial microwave system.

The overall capacity of communication satellites could be further increased through the use of SSB-AM that may allow as many as 12,000 voice circuits per transponder channel. However, the trend is toward digital time-division multiplexing of voice circuits with a reduction in overall capacity for a given transmission bandwidth. This is because the digital speech signal of 64,000 bps has a bandwidth of about 32,000 Hz, even though the bandwidth of the analog speech signal is only 4,000 Hz.

The capacity of communication satellites for carrying data is quite impressive. As an example, *Comstar IV* has 24 transponder channels, each capable of carrying 670 Mbps. The overall capacity of the satellite thus is 1.4 Gbps. This equates to transmitting all the text in 1.1 million 200-page books in one hour. Although this may seem impressive—and indeed is—optical fiber has far greater capacities and avoids the problem of delay. Optical fiber is described in the next chapter.

Satellites—an assessment

The delay problem inherent in geostationary satellite transmission is a serious disadvantage for interpersonal telephone conversation. One way of minimizing the problem is to use a satellite circuit for transmission in one direction and a terrestrial circuit for the return transmission. Satellites simply are not that suitable for voice transmission unless the transmission is to a very distant place that is otherwise unreachable.

Communication satellites do offer considerable bandwidth at a transmission cost that is independent of distance. They are quite suitable for the one-way broadcast transmission of television signals between the various television network locations and from the network to affiliated stations. Nearly all network and cable television broadcasters rely heavily on satellites for the distribution of program material.

A single satellite could broadcast television directly to our homes. A direct broadcast satellite (DBS) utilizing the 12-GHz band (12.2 to 12.7 GHz) transmits a signal that can be received by a small two-foot antenna. The technology for DBS is at hand, and a number of systems are currently serving the United States using digital transmission with compression of the TV signal to conserve bandwidth. However, one wonders whether there is much more of a market for more sources of television programs given the widespread penetration of cable television and VCR machines.

Earth station operating in the Ku band can have small antennas, thereby increasing the usefulness of satellite communication for data and voice applications. Such small antenna systems are known as very small aperture terminals (VSATs). The antenna typically is from 0.75 to 1.8m in diameter. Some VSATs are the size of a laptop computer and can be used for voice communication to reach areas without any other form of telephone or wireless communication.

Satellite technology continues to advance. Rocketry and space technology have advanced so that large payloads can be placed in orbit quite economically. Competition in rocket launching from private industry, as advocated by some people, could lower launch costs even more. Satellites need to carry fuel for orbiting repositioning. Ion beams could accomplish this repositioning using electric power obtained from solar panels. The amplification of the signal back to earth on the satellite is performed by traveling wave tubes (TWTs). The lifetime and reliability of these tubes continues to improve, thereby extending the useful life of the satellite. The present 4-deg spacing will decrease to 2 deg so that more satellites can share the same geostationary orbit.

Four-wire-to-two-wire conversion

The local loop is a two-wire circuit. As shown in Figure 4.17, the induction coil in the antisidetone circuit in the telephone instrument separates the signal generated by the transmitter from the signal intended for the

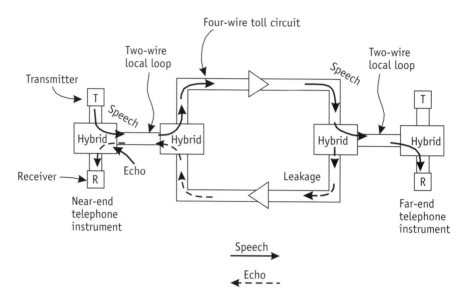

Figure 4.17 Hybrids match the two-wire local loop to the four-wire circuits used in toll transmission. The hybrid cannot perfectly match the impedance of every possible local loop, and hence some portion of the signal leaks through the hybrid and returns down the four-wire circuit in the opposite direction. Furthermore, electrical reflections occur along the local loop, and these reflected signals too are passed back down the four-wire circuit along with the signal that leaks through the hybrid. The returned signal is heard as an echo.

receiver. The two-wire circuit from the transmitter and the two-wire circuit to the receiver constitute a four-wire circuit. The induction coil thus converts a two-wire circuit into a four-wire circuit. Another name for the induction coil, or transformer, that performs such conversion is a hybrid.

Long-distance and toll circuits are unidirectional requiring separate paths for the signals traveling in each direction. This is because the amplifiers and repeaters used in long-distance and most interoffice-trunk carrier systems are inherently unidirectional. These circuits thus are four-wire circuits, and they must be matched to the two-wire circuits used in the local loop. This matching is accomplished by hybrids. A hybrid is required at each end of a toll circuit. The hybrid at the end near the speaking party is called the near-end hybrid; the hybrid at the other end is called the far-end hybrid.

Each hybrid used in the toll network consists of two transformers with taps on the primary and the secondary. A far-end hybrid is shown in Figure 4.18. A balance circuit attempts to match the impedance of the two-wire local loop. Consider a signal coming into the hybrid from the four-wire circuit. This signal passes through the primary of the in-coming transformer generating identical voltages at the two halves of the tapped secondary. One of these voltages generates a current that passes down the local loop and returns through one half of the tapped primary of the out-going transformer. The other voltage generates a nearly identical current that flows through the balance circuit and returns through the other half of the tapped primary of the outgoing transformer. The reversed leads of the balance circuit mean that the currents flowing through each half of the primary of the outgoing transformer are out of phase with respect to each other. The voltages induced at the secondaries are also out of phase and cancel. Thus, no current flows from the out-going transformer to the four-wire circuit.

If hybrids were ideal, they would perform perfectly, and no signals would leak through them to return down the four-wire circuit. To

Figure 4.18 The hybrid is constructed from transformers. A signal entering the far-end hybrid from the four-wire circuit appears only at the two-wire circuit and does not return through the four-wire circuit.

perform perfectly, the two currents in the primary of the out-going transformer must match perfectly. For this to occur, the matching circuit must match perfectly the impedance of the local loop. Since each local loop has a different impedance depending upon its length and other factors, such perfect impedance matching is not possible. Thus, some current will leak across the hybrid, and some of the input signal will be returned back down the four-wire circuit toward the near end and will be heard by the speaking party. If the four-wire circuit is long, the returned signal will arrive with a delay and will be heard as a distinct echo by the speaking party.

If the impedances are not matched perfectly at various discontinuities along a transmission line, electrical reflection of a signal traveling down the line will occur. Discontinuities occur at electrical connections, and there are usually a number of them along a local loop. Thus, a number of small reflections occur for a signal traveling along a local loop, and these reflections travel back the loop and appear as additional echo to the speaking party.

The returned signal can leak through the near-end hybrid and travel along the four-wire circuit to the far-end hybrid where it could leak again, and so forth. The net effect is to give the echo a reverberant quality, as if the person were speaking in a barrel. All this is most annoying, and for some amounts of delay, normal speech is impossible. But there is a solution to the problem of echo.

Echo elimination

Echo can be a serious problem on toll circuits, and, therefore it must be eliminated. One solution is to break or open the return path in the four-wire circuit so that the echo signal cannot return to the near end. This is accomplished by devices called echo suppressors.

As illustrated in Figure 4.19, an echo suppressor prevents the echo from returning down the four-wire circuit by either completely opening the return circuit or by inserting sufficient loss in the return circuit to reduce the echo signal to an inaudible level. This prevents echo caused by either leakage through the hybrid or reflections along the local loop. A control circuit senses when a signal is at the input to the hybrid and then activates the loss in the return path to prevent any echo. A compare

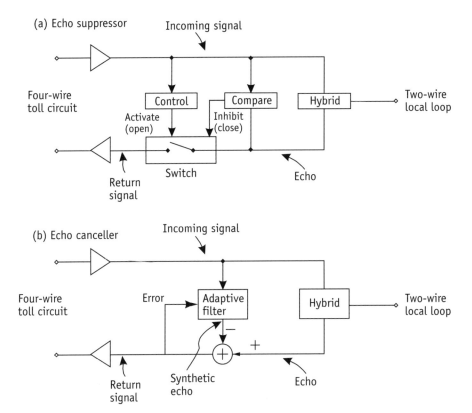

Figure 4.19 (a) The echo suppressor opens the return path and thus prevents echo. Unfortunately, it also prevents the speech from the other party from being heard during double talking. (b) The echo canceler uses a dynamic adaptive filter to calculate and subtract a synthetic echo from the return signal. It allows double talking.

circuit compares the input to the hybrid with its output to determine when the loss should be removed from the return path.

Early echo suppressors completely opened the return circuit. This meant that the other party's speech could never be heard during periods of double talking, and the circuit became half-duplex. Delay was deliberately designed into the dynamics of the control and compare circuits so that abrupt changes would not occur during the normal silent pauses in human speech. What all this meant was that unless the speaking party stopped talking entirely for enough time for the return circuit to be closed, the other party could not capture the return circuit.

Improved echo suppressors insert loss in the return circuit, which allows some of the other party's speech to be heard during double talking. But echo suppressors with their many adjustments, switching delays, and other problems are far from ideal. The problems of echo suppressors are avoided with echo cancelers.

Echo cancelers create a synthetic echo that is subtracted from the return signal thereby canceling the echo, as shown in Figure 4.19. Because different local loops have different electrical characteristics, the precise echo cannot be known in advance. The cancellation is therefore performed with electronic filters that can change their properties dynamically to minimize any portion of the returned signal that resembles the incoming speech signal. Echo cancelers have become affordable through the use of integrated circuits and digital processing. A short amount of time is needed by them to adapt to the properties of the connection, and a careful listener can hear the quick disappearance of the echo at the very beginning of the call. Echo cancelers allow two-way talking and full-duplex operation on long-distance calls.

Transmission impairments

The quality of transmission can be impaired by a variety of sources. These sources all produce unwanted signals that can be grouped under the general category of noise.

Amplifiers generate noise in the form of hiss. Because the frequency spectrum of this noise is uniform with all frequencies equally present, like the spectrum of white light, such noise is called white noise by analogy. Motors turning on and off, electromagnetic disturbances produced by lightning, and other such sources can generate clicks and pops. These clicks and pops are called impulse noise, since the noise looks like sharp spikes or impulses. Loose connections can produce a sputtering noise. The carbon microphone is a source of granular noise. The quantization used with digital conversion of the speech signal results in a small error upon reconstruction of the analog signal, and this error appears in the form of quantization noise.

Crosstalk is the leakage of one signal into another circuit. It can occur in the local loop where many lines all run parallel to each other. Crosstalk can also occur on long-distance circuits using frequency-division multiplexing when the signal from one channel leaks into another because the frequency shifting drifts a little or because of other causes.

Transmission systems cannot be perfect in the sense of eliminating all forms of impairments—that would be far too costly. Instead, the design goal is to reduce their subjective effect on users to an acceptable level. Hence, the human factors of transmission impairments are an important consideration in the performance specifications for transmission systems.

Selected bibliography

"AR6A Single-Sideband Microwave Radio System," *Bell System Technical Journal*, Vol. 62, No. 10, Part 3, Dec. 1983.

Elbert, B. R., *Introduction to Satellite Communication*, Norwood, MA: Artech House, Inc., 1987.

Hudson, H., *Communication Satellites*, New York, NY: The Free Press, 1990.

Keiser, B. E., and E. Strange, *Digital Telephony and Network Integration*, New York, NY: Van Nostrand Reinhold Company, 1985.

O'Neill, E. F. (ed.), *A History of Engineering and Science in the Bell System: Transmission Technology (1925-1975)*, Indianapolis, IN: AT&T Bell Laboratories, 1985.

Talley, D., *Basic Carrier Telephony*, Rochelle Park, NJ: Hayden, 1977.

5

Digital Transmission Systems

Introduction

Indeed, the world of telecommunication has gone digital. Switching machines utilize time-division switching of digital signals, and transmission systems utilize time-division multiplexing of digital signals. This chapter describes digital multiplexing in general and discusses the use of optical fiber as a transmission medium to carry digital signals.

Digital multiplexing

Time-division multiplexing is used to combine a number of digital signals. A sample of each digital signal is given its own unique time interval or time slot, and a large number of time slots are then transmitted repetitively in this fashion. The digital signals can be either an analog signal (such as a telephone speech, facsimile, or modem signal) that has been converted into a digital format or digital data directly generated by a computer.

The first step in converting an analog speech signal into a digital format is to filter the signal to prevent false, or alias, frequencies from appearing later upon reconstruction of the analog signal. The telephone speech signal therefore is lowpass-filtered to 4 kHz. The next step is to sample the band-limited signal at the Nyquist rate of 8,000 samples per second. The corresponding sampling interval is 1/8,000 sec, or 125 μs.

In the world of telephony, the sampled signal is then quantized into either 256 or 128 levels, depending upon the number of bits that will next be used to encode each quantized sample. Eight bits are used with 256 levels; seven bits are used with 128 levels. Usually, eight bits per sample are used. The bit rate corresponding to eight bits per sample is 64,000 bits per second. A 64-kbps digital version of a single voice signal is called a DS-0 signal. Since the amplitudes of the signal are encoded as binary combinations of on-off pulses, digital is also known as pulse code modulation.

The sampled signal is usually compressed in amplitude and then quantized using a linear quantizer having equal-sized quantization steps. This allows more resolution or steps for smaller amplitudes of the signal since speech signals rarely have very large amplitudes. The reconstructed signal will be later expanded upon reconstruction to obtain the original dynamic range. The process of compression followed by expansion is called companding. The compression characteristic is logarithmic, with more compression as the amplitude of the signal increases. The μ- law is used in North America and Japan, while the rest of the world uses the A-law. Conversion is thus required on international calls between these two parts of the world.

Many digital signals are time-division multiplexed together to create the North American asynchronous digital hierarchy. Time-division multiplexing of 24 voice channels gives a DS-1 signal, which requires a bit or data rate of 1.544 million bits per second (Mbps). The 24 voice channels require 1.536 Mbps and an additional 8 kbps are needed for synchronization purposes, thereby giving 1.544 Mbps as the overall bit rate. A group of 24 voice channels digitally multiplexed together is sometimes called a digital group, or a digroup for short.

As shown in Figure 5.1, four DS-1 signals time multiplexed together give a DS-2 signal containing 96 voice channels and requiring a data rate of

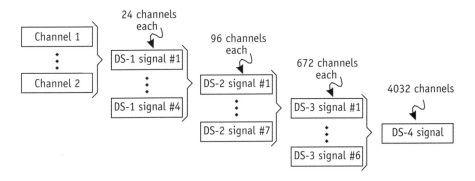

Figure 5.1 A D-type channel bank combines voice signals through time-division multiplexing. The basic multiplexing unit consists of 24 voice channels, sometimes called a "digroup." A single voice channel is digitized at 64,000 bits per second and is sometimes referred to as a DS-0 signal. A digroup, or 24 multiplexed voice signals, is called a DS-1 signal.

6.312 Mbps. Seven DS-2 signals time multiplexed together give a DS-3 signal containing 672 channels and requiring 44.736 Mbps. Six DS-3 signals time multiplexed together give a DS-4 signal containing 4032 channels and requiring a data rate of 274.176 Mbps. The extra bits in the higher capacity digital signals are used for timing and synchronization information to assist in the separation and demultiplexing of the individual channels.

The timing and clocking information is contained within the digital bit stream and thus is self-synchronizing, or asynchronous. Such near synchronous timing schemes form a plesiochronous digital hierarchy (PDH).

Modulation and multiplexing

Multiplexing is the combining together of a number of signals to share a transmission medium. Analog modulation is the shifting of signals in frequency through either the amplitude modulation or the frequency modulation of a pure-tone carrier wave. Digital modulation is the sampling and encoding of an analog signal as a series of binary pulses.

The world of transmission has gone digital, but that does not mean that analog modulation is dead. Most transmission media carry such analog waves as light or radio signals. This means that although time-division multiplexing is used to multiplex together a number of digital signals, the multiplex signal must then be transmitted over the transmission medium using analog modulation of a carrier wave. For example, thousands of voice channels are time-division multiplexed together for transmission over optical fiber. But the digital multiplex signal ultimately is used to turn on and off the analog wave of light that is transmitted along the fiber—a form of analog amplitude modulation. As another example, the digital multiplex signal transmitted over microwave radio ultimately must modulate an analog radiofrequency carrier wave that shifts the digital multiplex signal into the frequency band of the assigned microwave radio channel.

Most transmission media ultimately carry analog waves. But these analog waves might represent digital multiplex signals that contain thousands of speech signals that have been time-division multiplexed together. The analog wave is varied in amplitude, frequency, or phase in synchrony with the multiplex signal. The one exception is baseband transmission of a digital signal, in which the actual multiplex signal is carried directly as an electric signal over the medium, usually copper wire in the form of twisted pair or coaxial cable.

T1 digital carrier

Basic principles

Digital carrier systems utilize time-division multiplexing to combine a number of digital signals. The digital microwave radio systems previously described are examples of digital carrier systems. Other transmission media can be used to carry time-division multiplexed digital signals. The digital signals are usually speech signals encoded at 64,000 bps, but can also be data transmitted between computers and computer terminals. Voice telephony generates considerably more traffic than other sources, and hence most of the traffic carried in digital form is from voice circuits.

The earliest use of digital carrier, or pulse code modulation (PCM), was the T1 system, which first saw service in 1962. The T1 digital system is used on short-haul interoffice trunks less than 50 miles long. Conventional 16- to 26-gauge copper wire pairs are used, with a pair needed for each direction. The digital information is transmitted directly as a baseband electrical signal over the wire. With distance, the on-off pulses that form the baseband digital signal become smeared together. Before it is no longer possible to distinguish one pulse from another, the pulses are detected and are regenerated again for transmission over another length of the wire pair. This process is performed by electronic devices, called regenerative repeaters, that are located every mile along the wire pair.

Twenty-four digital voice circuits are time-division multiplexed together in the T1 system. Each digital signal consists of a series of 8-bit samples. The 24 digital signals multiplexed together thus become a string of 8 × 24 or 192 bits, called a frame, as shown in Figure 5.2. The demultiplexing equipment at each end must be very precise in decoding the fast string of bits so that the 8 bits for each circuit are assigned and decoded correctly. To allow recovery from any mishap, an extra bit is assigned to each string of 192 bits. This 193rd bit alternates from on to off each frame. This regular pattern of bit alternation can be used by the demultiplexing equipment to identify the 193rd bit and thus regain synchronization. The 193rd bit is therefore called the synchronization bit.

All 193 bits that form a frame must be transmitted in 1/8,000 sec, or else the bits from the incoming signal will start to accumulate. Thus, the overall bit rate is 193 bits per 1/8,000 sec, or 1.544 Mbps. Each individual bit must be sent in 1/1,544,000 second, or 0.65 microsecond (µs).

Figure 5.2 In the T1 digital carrier system, 24 digital circuits are combined together. Each digital circuit contains 64,000 bits per second, corresponding to the analog speech signal that has been sampled 8,000 times per second with 8 bits being used to encode each sample.

One of the eight bits in each digital voice channel is used every sixth frame for signaling and supervisory purposes, for example, to specify that a particular channel is idle or in use.

Bipolar pulses

The digital information in the T1 system is transmitted in the form of bipolar pulses. A binary 0 is encoded as zero volts. A binary 1 is encoded alternately as positive pulses or as a negative pulse, as shown in Figure 5.3. These pulses have a width equal to one-half the bit time of 0.65 µs. The alternating polarity of the pulses gives an average value of zero to the digital signal and also makes it simple to detect a single-bit error.

If conventional polar pulses with a width of 0.65 µs were used, the maximum frequency of a pulse train with alternating 0's and 1's would be about 772 kHz. The use of pulses with half that width would normally double the fundamental frequency and hence the bandwidth needed to pass that signal. However, the bipolar format has its maximum fundamental frequency for a string of binary 1's, and that frequency is also about 772 kHz. Thus, the bandwidth requirements are not increased by the use of bipolar pulses. However, because there are three voltage levels with the bipolar format, a decision based on two thresholds must be made at the receiver to determine the binary information. This is more complicated than with the simpler polar format, where a decision based on only one threshold must be made because there are only two possible voltage levels.

Expanded systems

As we saw previously with digital microwave radio, a number of DS-1 digital signals can be multiplexed together and transmitted over any medium. The bandwidth of the medium will determine the overall rate and capacity that the medium can support.

Conventional wire pairs (twisted pair) are used as the transmission medium for the T1 system. Special 22-gauge wire pairs with low electrical capacitance are used to create a digital carrier system with more circuit capacity than the conventional T1 system. This is because wire pairs with low electrical capacitance have a greater bandwidth than

(a) Bipolar format:

(b) Fundamental frequency:

(c) Thresholds:

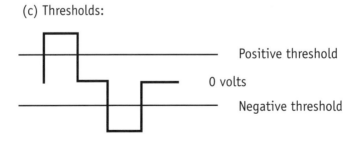

Figure 5.3   (a)The bipolar format used in the T1 system is a return-to-zero scheme in which a binary 0 is encoded as zero volts, and a binary 1 is encoded as a return-to-zero pulse that alternates in polarity from pulse to pulse; (b) even though the bipolar pulses are half the width of polar pulses, the maximum fundamental frequency of a string of pulses is the same; (c) the bipolar format requires two threshold decisions at the receiver.

conventional wires and thus can pass a higher data rate. This system is called the T2 digital carrier system.

Four DS-1 signals are time-division multiplexed together to create the DS-2 signal that is used in the T2 system. T2 was first introduced into the Bell System in 1972, for use on intercity trunks up to 500 miles in length. Regenerative repeaters are spaced about every three miles. Special low-capacitance 22-gauge wire is used. A total of 96 digital voice circuits are carried on each wire pair in a T2 system. Usually, 24 wire pairs are used in a cable, with one wire pair reserved as protection. The total capacity of a system is 2,208 two-way digital voice circuits, using a separate cable in each direction. Each digital T2 line operates at a bit rate of 6.312 Mbps. The bipolar format is used.

In the early 1980s, AT&T introduced an improved T-carrier system, called T1G, operating on standard twisted-pair cables. The T1G system carries 96 digital voice circuits in each direction over two twisted pairs. Rather than the bipolar pulses used in the older T1 system, the newer T1G system uses four-level pulses, called a quaternary format. Each one of the four possible levels represents one combination of two bits. For example, a voltage level of +3V represents the binary digits 11; +1V represents 01; −1V represents 00; and −3V represents 10.

Digital carrier is also used on coaxial cable. Coaxial cable with its much higher bandwidth than conventional wire provides for even higher data rates. The T4M digital carrier system uses coaxial cable and operates at 274.176 Mbps, corresponding to 4,032 digital voice circuits. A T4M system uses the same coaxial cable consisting of 11 pairs of coaxials used in the L5 system. One pair of coaxials is reserved for protection. The total capacity of the T4M system is 40,320 two-way digital voice circuits. The T4M system typically has been used on routes of 500 miles or less in length. A polar format is used with the T4M system to provide increased noise immunity compared to the bipolar format used in the T1 and T2 systems.

Digital microwave radio

In the early 1980s, AT&T introduced digital radio systems that used time-division multiplexing and operated in the microwave radio bands for use in short- to medium-length routes. The DR6-30 system was first used in 1981 and operates in the 6-GHz band with channels that have a bandwidth of 30 MHz. A total of 1,344 two-way digital voice circuits are transmitted in each of seven two-way radio channels. This corresponds to two DS-3 signals per radio channel at a rate of about 90 Mbps. The total system capacity using all seven radio channels is 9,408 two-way digital voice circuits. The DR6-30 system is used for hops from 15 to 30 miles.

At about the same time, AT&T introduced the DR11-40 system operating in the 11-GHz band with 40-MHz channels. The ten two-way radio channels carry a total of 13,444 two-way digital voice circuits. The higher frequency 11-GHz band is more susceptible to fading caused by rain, and hence the DR11-40 system is restricted to hops of 15 miles or less.

The time-division multiplexed signal is transmitted over the air using a technique called quadrature amplitude modulation (QAM). The digital signal is transmitted as a continuous series of short bursts of a radiofrequency wave. Each short burst, called a baud, can vary both in maximum amplitude and in phase. A total of 16 combinations of amplitude and phase are possible for each baud, as shown in Figure 5.4, and thus each baud can encode four bits. This specific type of quadrature amplitude modulation is called 16-QAM.

The challenge with QAM is the ability to distinguish reliably one combination of amplitude and phase from another in order to know precisely what information was transmitted. noise, fading, and interference all conspire to make this task difficult. Nevertheless, AT&T improved the technology and developed 64-QAM, which encodes and transmits six bits per baud with an error rate of less than 1 in 10 billion bits.

The DR6-30 system operating in the 6-GHz band and utilizing 64-QAM was introduced by AT&T in 1984. It transmits three DS-3 digital signals in each radio channel at a rate of about 135 Mbps. The total system capacity of the DR6-30-135 system is 14,112 two-way digital voice circuits, and the system replaces frequency-division multiplexing systems in the long-distance network. A DR11-40-135 system operating in the 11-GHz band with a total system capacity of 20,160 two-way digital voice circuits is also available.

Digital radio systems can operate in the 4-GHz band with further increases in system capacity. The DR4-20-90 system operating in the 4-GHz band is capable of carrying two 45-Mbps signals per radio channel.

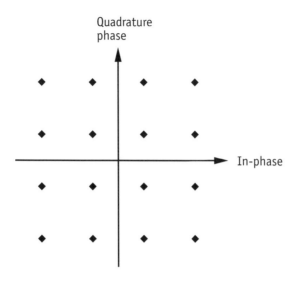

Figure 5.4   Each short burst of the radio wave can have one of 16 possible combinations of amplitude and phase. The specific combination determines the four bits that were transmitted for that short burst. This type of digital transmission is called quadrature amplitude modulation (QAM).

With 11 active two-way radio channels, the DR4-20-90 system is able to carry 14,784 two-way digital voice circuits.

The older DR18 digital microwave radio system operating in the 18-GHz band has been used on very short hops of from one to five miles. This is because of attenuation of the radio signal caused by moisture in the air. Each channel in the microwave band operates at a 274.176-Mbps rate, corresponding to 4,032 digital voice circuits. A total of eight two-way radio channels are used with an additional two-way channel as protection for a total system capacity of 28,224 two-way digital voice circuits. Differential phase-shift keying (DPSK) with dibit coding is used to modulate the radio carrier in each channel.

Optical fiber

Basic principles

Imagine a very long mirror that has been rolled up to create a long pipe with the mirrored surface on the inside of the tube. Alternatively, imagine a long pipe with an inside surface that is mirrored. If a source of light shines into one end of the tube or pipe, the light waves will be repeatedly reflected off the surface of the mirror and guided along the length of the pipe until they emerge at the other end. In effect, the mirrored tube is a light, or optical, waveguide. The use of mirrored pipes to carry light is an old idea: the American William Wheeler was granted a patent in 1881 on the use of mirrored pipes for guiding light.

There is a better way of creating a mirror to reflect light at shallow angles. An optical boundary is formed when two materials having different angles of refraction are joined. A light ray passing across this boundary is bent when it emerges in the second material, as shown in Figure 5.5. The angle at which the light ray is bent is proportional to the relative indices of refraction of the two materials—a principle known as Snell's law of refraction. If the light ray impinges on the boundary at a very shallow angle, called the critical angle, it will be reflected at the boundary as if the boundary were a perfect mirror.

Consider a long, thin strand of glass in the form of a fiber. The interior core of the fiber is surrounded by a cladding of glass with a lower index of refraction than the core. If a light wave enters the core at a shallow angle, it will be repeatedly reflected at the boundary of the core and the cladding until it emerges from the end. This is the basic principle of optical fiber. It is a medium for transmitting light waves. It too is an old idea: the use of thin glass fibers for guiding light was described in 1887 by the British physicist Charles Vernon Boys.

The idea of using light as a transmission medium for speech signals was conceived in 1880 by Alexander Graham Bell with his invention of the photophone. It transmitted voice waves over the air on light waves. The synthesis of Bell's ideas with the light-carrying capabilities of thin

Figure 5.5 (a) A light ray passing across an optical boundary is bent; (b) if the angle of incidence is less than some critical angle, the light ray is reflected at the boundary.

glass fibers did not occur then. It was not until quite recently that optical fibers have come into widespread use for communication. Digital multiplexing was not known a century ago, nor was the required electronic circuitry available. But optical fiber is indeed the transmission medium of choice today for telecommunication circuits across continents, under oceans, and between central offices. Someday, optical fiber may also be the choice for local circuits all the way from central offices to homes and offices.

The exterior of an optical fiber is coated with a jacket (see Figure 5.6) to prevent stray light from entering the fiber. The actual glass from which the optical fiber is manufactured is silica with extremely high purity. Typical losses are about 0.2 dB/km, and fibers with even lower losses are under development as the technology continues to progress. As a comparison, the very best optical glass has a loss in the order of about 1,000 dB/km.

An optical-fiber communication system consists of a light source, the fiber itself, and a light detector at the distant end, as shown in Figure 5.7. The light pulses are detected and regenerated along the length of the system at regular intervals, determined by the degree of smearing, noise, and losses in the fiber. The regeneration of pulses along the length of the system is called repeatering and is performed by electronic devices called regenerative repeaters.

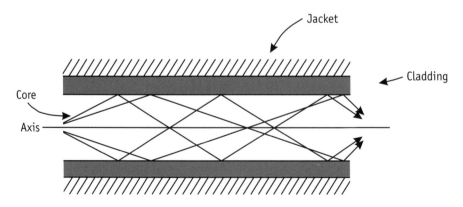

Figure 5.6   Light waves entering an optical fiber at less than the critical angle are continuously reflected through the glass core of the fiber.

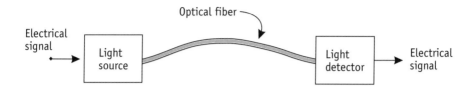

Figure 5.7   The light from a source is guided through the optical fiber to a detector at the distant end. This forms a basic fiber communication system.

Light is the carrier wave for signals multiplexed together and transmitted over optical fiber. Today, it is difficult to modulate the amplitude or frequency of light in a highly linear fashion. But what is very simple to do is to turn the light on and off. Optical fibers are thus particularly well suited for carrying digital signals that have been time-division multiplexed together. Light has an extremely high frequency and can be turned on and off at an extremely fast rate, and hence a very large amount of information can be carried over a single optical fiber. This large capacity combined with their freedom from noise and interference, low cost, low loss, and small size are some of the reasons why optical fibers have truly revolutionized transmission technology in such an exciting way.

Types of optical fiber

Light waves traveling through an optical fiber take different paths depending on their specific reflections. All these paths have different lengths, with the shortest path being right down along the axis of the fiber. These different paths mean that the light will arrive at slightly different times and will thus be smeared in shape (see Figure 5.8). The amount of smearing, or pulse spreading, determines the rate at which data can be sent over the fiber. Some types of optical fiber are better than others in terms of smearing.

Three types of optical fiber are defined by optical grading of the fiber itself and the modes that the light travels during transmission. The concept of mode is concerned with the number of possible light paths through the fiber. Multiple-mode, or multimode, fiber has many possible paths along which the light travels and thus is subject to the most smearing.

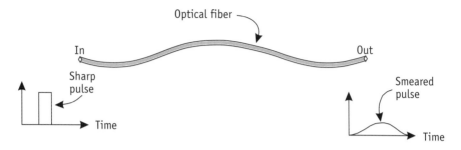

Figure 5.8   Because the light travels through the fiber on many paths of different lengths, a sharp pulse is smeared in time when it leaves the fiber. Losses in the fiber will attenuate the amplitude of the pulse. These effects limit the maximum capacity of the fiber.

The simplest type of multimode fiber consists of an inner core with a uniform index of refraction surrounded by an outer jacket with a lower index of refraction. This type of optical fiber, illustrated in Figure 5.9, is called multimode stepped-index fiber. The inner core is about 50 μm in diameter, or roughly the diameter of a human hair. If a sharp pulse of light is sent along the fiber, it emerges quite smeared because of the many different paths it travels. This smearing limits the maximum pulse rate to about two million pulses per second and the practical distance between repeaters to about five kilometers.

Light travels in a multimode manner through the second type of optical fiber, but the refractive index decreases gradually away from the axis of the core. The result of this graded index is that light rays are gradually turned back toward the axis of the core as they travel through the fiber. This type of fiber, illustrated in Figure 5.10, is called multimode graded-index fiber. Another way of visualizing the workings of multimode graded-index fiber is that the light rays are continuously refocused as if the fiber were a continuous succession of small lenses. The diameter of the core of multimode graded-index fiber is also about 50 mm. Because of the continuous refocusing of light rays, smearing of pulse is much less. The practical pulse rate for multimode graded-index fiber is about 100 million pulses per second.

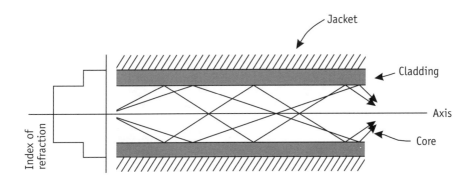

Figure 5.9 Refractive index and some representative light paths in a multimode stepped-index optical fiber.

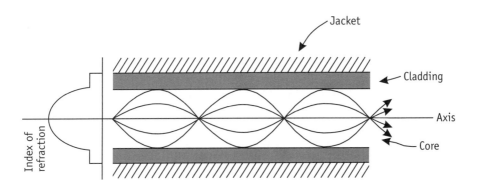

Figure 5.10 Refractive index and some representative light paths in a multimode graded-index optical fiber. The light is continually refocused as it travels along the fiber.

The third type of optical fiber is called single-mode fiber (see Figure 5.11). The diameter of the core of single-mode fiber is very small, typically about 5 μm. Only parallel rays of light can pass through by traveling along its axis, and multiple paths with their inherent smearing are avoided. More will be said later in this section about the capabilities of optical fiber.

Figure 5.11 Single-mode optical fiber has a core that is so thin that only parallel rays of light travel along it. In this way, all the rays of light arrive at the end at the same time, and smearing is nonexistent.

Source and detector

If the speed of an optical fiber system is less than 100 Mbps, then a single light-emitting diode (LED) can be used as the light source in an optical fiber communication system. LEDs are economical and very reliable.

If the speed is more than 100 Mbps, then a laser must be used as the light source. The light output of a laser is a collimated beam of single-frequency coherent light. The collimated beam of parallel light rays makes it easy to couple the beam to a single-mode fiber. Lasers used to be large and costly with shorter lifespans than LEDs. However, all that changed with the advent of solid-state miniaturized lasers, describer later in this chapter.

A light-sensitive semiconductor diode is usually used as the light detector. Semiconducting material is photosensitive, and its resistance varies with the intensity and frequency of the light stimulating it. The resistance decreases as the illuminating light increases in intensity. A semiconducting diode is used as a photodiode by allowing light to illuminate the semiconducting material at or near the depletion region by the junction. The diode is reverse biased, and the leakage current varies according to the light absorbed at the deletion region. Photosensitive diodes are sometimes made from a sandwich of positive-doped, intrinsic (or natural), and negative-doped semiconducting materials. Such diodes are called PIN diodes.

If the reverse bias is increased to a high voltage, the electrons freed by the light will collide with the other atoms and will have sufficient energy to create hole-electron pairs which, in turn, will collide with other atoms, creating more hole-electron pairs, and so on in an avalanching chain-reaction fashion. Such diodes are thus called avalanche diodes. The effects of the multiplication of the current carriers give these devices an internal gain and an increased sensitivity. Avalanche diodes are also used as photodiode detectors in optical-fiber communication systems.

Lasers

A single-mode optical-fiber communication system uses a laser as its light source, and hence we should understand the principles of operation of lasers. A laser generates a beam of light with all the light rays parallel to each other. This property of laser light is called collimated. Furthermore, the light rays are all in phase with each other, a property called coherent. Lastly, the light usually is a single frequency, a property called monochromatic. Lasers generate a collimated beam of coherent monochromatic light.

The principles of operation of lasers involve quantum mechanics. (See Figure 5.12.) If an atom is excited by an external source of energy, the atom will absorb some of the energy and some electrons will move into higher energy bands. The atom is not stable in this excited state and will return to its normal unexcited state rather quickly. When this happens, the excited higher energy level atoms fall back to their normal levels. In doing so, they release precise quanta of energy in the form of light at a precise frequency, or a set of precise frequencies, depending on the particular type of atom. This release of energy in the form of light is called spontaneous emission. It is the way fluorescent lamps work.

The emission of light energy can also be stimulated by further exciting an already excited electron with additional light energy at precisely the correct frequency. This form of light emission is called stimulated emission and forms the basis of operation of a laser.

A laser consists of a chamber, called a resonator (or resonant chamber), containing a material whose atoms have the property that they can be excited to produce laser light (see Figure 5.13). The material is said to be able to lase and could be either a solid or a gas. A mirrored surface is

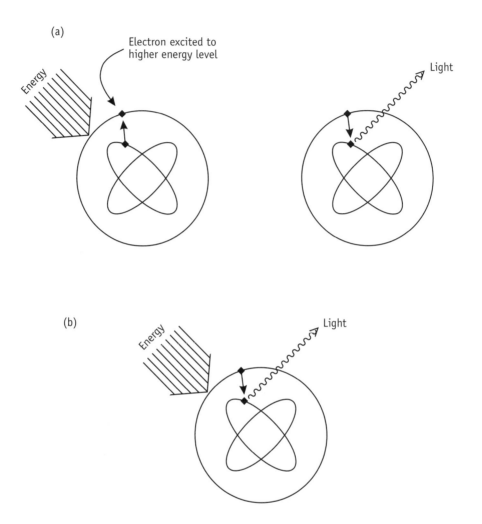

Figure 5.12 (a) Spontaneous emission of light occurs when an electron is excited to a higher energy level by an external source of energy and then returns spontaneously from this unstable level to its normal, stable level; (b) an electron already excited to a higher state can be further excited by external energy to return to its stable state.

placed at each end of the two parallel ends of the chamber. An external source of energy is then applied to the material in the chamber. This causes the atoms to be excited, and some of them emit energy in the form of the spontaneous emission of light. The light energy thus produced

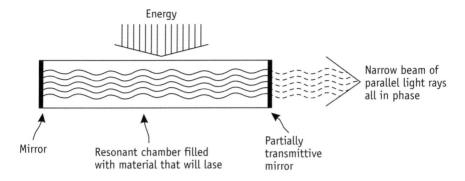

Figure 5.13 In a laser, external energy produces the stimulated emission of light. The light is reflected from mirrors at each end of a resonant chamber, thereby creating an optical standing wave that stimulates the emission of more light. A portion of the light passes through a partially transmittive mirror at one end of the chamber. The laser light is a monochromatic, coherent, collimated beam of light.

from spontaneous emission then stimulates adjacent already excited atoms to emit light through stimulated emission. This light then travels to other adjacent atoms, causing the further stimulated emission of light in a cascade fashion. Quickly, a wave of light travels through the material in the chamber until the wave encounters one of the mirrors.

When the mirror is encountered, the light wave is reflected and travels back through the material in the chamber, thereby stimulating the emission of more light energy. The mirror at the other end is encountered, the light is reflected again, and the release of more light is stimulated in the chamber. The continuously reflecting wave of light quickly reaches a fairly high intensity, and the process continues as long as the external source of energy is applied. The distance between the mirrors is chosen to create an optical resonant chamber, and therefore an optical standing wave is created in the chamber. One of the mirrored surfaces is partially transmissive, and thus some of the light can pass through the mirror and emerge from the chamber. The process involves Light Amplification by the Stimulated Emission of Radiation, hence the term **laser**.

Because there is one wave traveling back and forth in the resonant chamber, the light wave is coherent in phase. Also, the stimulated emission produces light of a single, precise frequency. That frequency, or color, is dependent on the atomic properties of the material in the chamber. Some materials will lase at more than one frequency. The laser light is generated by many internal reflections within the chamber, and hence the light rays are all parallel to each other. Because all the light is concentrated in a single beam, the beam of light can be quite intense. Laser light can be so intense that it can be used as a surgical knife and can even burn holes in metal.

Since not all atoms will lase, the material in the chamber must be the correct kind to produce laser light. The material can be a gas, a solid, or even a liquid. Some gaseous materials used in lasers are helium-neon, carbon dioxide, and argon. Ruby is one solid material that has been used in lasers. The semiconductor material used in a diode will lase, too. The small size and low power requirements of semiconductor diodes when used as lasers make them particularly well suited for consumer electronics and communication applications.

We are already familiar with the LED used for visual displays in clocks and other appliances. The type of light emitted by an LED is conventional light. However, if the diode is constructed of the proper semiconducting material with a junction in the form of a small resonant chamber, and if the junction is forward biased beyond a threshold current, the diode will emit laser light from the junction, as shown in Figure 5.14. Laser diodes usually use gallium arsenide (GaAs) as the semiconducting material. The junction forms the optical resonant chamber. One end of the junction is made optically reflective by a metallic coating, and the other end is partially reflective so that the laser light can escape. The other sides of the chamber are cut rough to inhibit lasing in the wrong direction. The threshold current is about 100 milliamperes (mA).

Erbium-doped optical amplifiers

Although optical fiber has a very low loss and little dispersion occurs in single-mode fiber, the light signal does decrease in intensity over distance and needs to be boosted along the way. This is done with regenerative repeaters. The light signal is detected and converted to an electrical signal. The electrical signal is then threshold analyzed to determine the presence of a pulse, and a new electrical pulse is then generated. The electrical signal is then finally converted back to a light signal to be sent along the

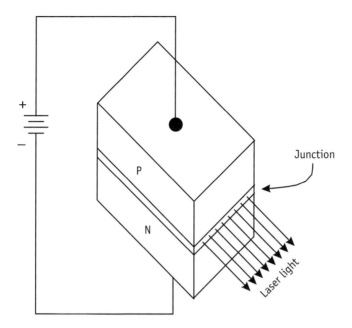

Figure 5.14   A semiconductor diode will emit laser light from its forward-biased junction. The junction is constructed in the form of a small optically resonant chamber.

next span of optical fiber. If only there were some magical way to amplify the light pulses without the need to convert them to electrical signals and back to light again. Erbium-doped fiber amplifiers perform that magic.

The basic principle of light amplification was discovered in the early 1960s by Charles J. Koester and Elias Snitzer of the American Optical Company, using rare earth neodymium. A short length of fiber, typically a fiber loop about 10m long, is doped with erbium. This fiber loop itself becomes a laser when it is excited by an external source of energy. When a pulse comes along, it kicks the energy into a higher level that then instantly falls back, creating a energized pulse. External power as small as 10 mW gives gains of 30–40 dBs.

Solitons

Different wavelengths of light travel at slightly different velocities through an optical fiber, the so-called optical Kerr effect. By carefully matching different wavelengths to different intensities, pulses called solitons are created. A soliton pulse travels at one speed, and hence no dispersion or compression of the pulse occurs as it travels along the fiber. The soliton pulse will decay in overall intensity, but it retains its shape. Soliton pulses have been transmitted over fibers as long as 10,000 km with little broadening in shape. Rates as high as 2.5 Gbps over single lengths of 14,000 km have been achieved using solitons.

Wavelength-division multiplexing (WDM)

Light sources centered at different wavelengths can be used on the same fiber with each source carrying different signals. In effect, the fiber is being multiplexed in the frequency domain. However, in fiber systems such frequency-division multiplexing is known as wavelength-division multiplexing (WDM). As many as five or more channels can share the same fiber. Wave-division multiplexing is clearly the way of the future since it allows substantial increases in the capacity of existing fiber.

Synchronous Optical Network (SONET)

The T1 system obtains its clocking information from the bit stream itself. Such a method is called asynchronous. An alternative method is the use of synchronized timing and clocking obtained from a highly accurate master clock. Such an approach to optical networking is called a synchronous optical network (SONET) and was first introduced in 1984. The SONET standard facilitates interconnection between optical networks operated by different common carriers and between different facilities within a network. With SONET, a single DS-1 signal can be extracted from the bit stream without the need to demultiplex the entire bit stream.

The basic signal supported by the SONET standard is called a synchronous transport signal at level 1 (STS-1). An STS-1 signal consists of a frame of 810 bytes (or 6,480 bits) transmitted in 125 ms, or an overall bit rate of 51.84 Mbps. STS-3 operates at 155.52 Mbps and is achieved by multiplexing three STS-1 signals together. The STS is also called the electrical signal. The optical carrier (OC) operates at STS rates and higher, and OC-1 and OC-3 are identical to STS-1 and STS-3. Other optical carrier operation levels are OC-12 at 622.08 Mbps, OC-24 at 1,244.16 Mbps, and OC-48 at 2,488.32 Mbps.

A synchronous digital hierarchy (SDH), similar in concept to the digital signal (DS-n) hierarchy, was developed for optical networks. The basic rate in the hierarchy—called a synchronous transfer module (STM-1)—consists of a frame of 2,430 bytes transmitted in 125 ms, or an overall rate of 155.52 Mbps. Other rates in the SDH are STM-4 operating at 622.08 Mbps and STM-16 operating at 2,488.32 Mbps.

Optical transmission facilities are installed in the form of rings—called SONET rings—so that there are two transmission paths to reach any destination in case of a network failure. Add-drop multiplexers (ADMs) and digital cross-connect systems (DCSs) complete a SONET network.

AT&T's domestic long-distance network in the late 1990s consisted of about 41,000 route miles of optical-fiber systems. Over 97 percent of AT&T's traffic was carried over fiber. About 2 percent of AT&T's traffic was still being carried over digital microwave in the late 1990s, but mostly in remote mountainous areas or to complete a network loop. A small amount of traffic was being carried over coaxial cable for secure government applications.

Capacity

The increase over time in the capacity of commercially available optical fiber communication systems is astonishing. AT&T's first fiber system, FT3, was introduced in 1979 and used graded-index fiber at 45 Mbps capable of carrying 672 voice circuits (one DS-3 channel). A total of 72 fiber pairs were placed together in a single cable, and 66 of the pairs were in service (with the remainder as spares). The total route capacity of the system was 44,352 two-way digital voice circuits.

The FT3C system was introduced soon afterward, with each fiber operating at 100 Mbps and carrying 1,344 digital voice circuits. Using cable with 66 active pairs, the system had a maximum route capacity of 88,704 two-way digital voice circuits. Repeaters were located every four miles along the route.

In 1984, AT&T introduced the FTX-180 system, with each single-mode fiber operating at 180 Mbps and carrying 2,688 digital voice circuits. A 400-Mbps system was introduced in 1986, with each single-mode fiber carrying 6,048 digital voice circuits and repeaters spaced every 20 miles. Fiber systems operating at 2,000 Mbps (2 Gbps) are standard.

The capacities of today's fiber systems are so mind-boggling that something other than Mbps needs to be used. I like to remember that 50 Mbps is equivalent to transmitting all the text in eight paperback books in one second, or carrying 780 digital voice circuits, or carrying one uncompressed color television signal. Using this method of equivalency, 2,000 Mbps is equivalent to 320 paperback books per second, 31,250 digital voice circuits, or 40 television signals. This is most impressive, but is nothing compared to the theoretical limit of single-mode fiber.

The theoretical capacity of a single light channel in single-mode fiber is about 200 Gbps—that is, 200 billion bits per second (200×10^9 bps). This capacity could carry 30,000 paperback books per second, 3 million digital voice circuits, or 4000 television signals. If the entire light spectrum operable for fiber (the 1.45- to 1.65-μm wavelength region) were used through color multiplexing, then the theoretical capacity increases to 50 Tbps, or 50,000 Gbps (50×10^{12} bps). This capacity could carry 7.5 million paperback books per second, 750 million digital voice circuits, or 1 million television signals. All these impressive capacities are theoretical maxima, and we are nowhere near such capacities, nor would we want to be. Clearly, it would make little sense to put all our signals in one fiber basket because, in case of trouble, the damage would simply be too great.

Optical fiber—an assessment

One advantage of optical fiber is its low loss, which means that very few repeaters are needed along a fiber route. For example, in August 1983, Bell Labs scientists gave a laboratory demonstration of an unrepeatered 100-mile length of fiber operating at 420 Mbps. Clearly, an unrepeatered strand under the Atlantic will someday be possible.

One problem with fiber was that the many vendors all had systems that did not interconnect. This meant that the optical signal had to be demultiplexed and then multiplexed again when two fiber systems were joined. The solution is an interface standard that enables the use of standardized rates operating synchronously so that the transmitter and receiver are locked together at a fixed rate of transmission. This standard is called SONET for Synchronous Optical Network. The basic SONET rate in the United States is 51.84 Mbps; in Europe it is 34 Mbps. Multiples of these rates, such as 155.52 Mbps, have been standardized, too.

Optical fiber offers many advantages over other transmission media. The fiber itself is flexible and inexpensive. Because light is carried by the fiber, optical fiber is not prone to electrical and electromagnetic interference. Because no light escapes from the fiber nor can enter it from outside, there is no leakage or crosstalk between adjacent fibers. The bandwidth and capacity are extremely high, and losses are quite small.

The extremely small size of fiber—one-tenth the diameter of a human hair for single-mode fiber—makes splicing very tricky, but techniques have been developed to speed that task. However, switching must be accomplished by converting the optical signal to an electrical signal and then using conventional techniques. Similarly, amplification and regenerative repeatering are accomplished by converting the light signal to an electrical signal. In both cases, it would perhaps be more efficient if these tasks could be accomplished directly on the light signal without the need for conversion to an electrical signal. Erbium-doped optical amplifiers solve some of these problems.

Single-mode fiber clearly is the winner. By using a number of different light signals at different frequencies, the overall capacity of a single strand of fiber can be increased—a technique called color multiplexing, or wave-division multiplexing. Semiconductor lasers are tiny and inexpensive, and are the obvious choice as light sources in fiber systems.

Transoceanic undersea cable

The challenge

Cable is laid under the sea for communication across oceans. Clearly, it is difficult and costly to make repairs of cable and repeaters located far under the ocean. Hence, reliability is the prime consideration in transoceanic undersea cable systems. This reliability is obtained by the use of conservative technology and by a trade-off against system capacity. Continual improvements in repeater design allowed closer spacing and more repeaters to be used, with a resulting increase in system capacity, as shown in Table 5.1.

Table 5.1
The History of Transatlantic Submarine Cable

System Name	Service Date	Cable		Repeaters			Two-way Voice Circuit Capacity		Digital Capacity
		Type	Technology	Quantity	Spacing Statute Miles	Technology	System Design	Maximum TASI	
TAT-1	1958*	SB	Two coaxials	51	44	Vacuum tubes	36	72	
TAT-2	1959*	SB	Two coaxials	57	44	Vacuum tubes	36	72	
TAT-3	1963*	SD	One coaxial (1.0 in)	182	23	Vacuum tubes	138	276	
TAT-4	1965*	SD	One coaxial (1.0 in)	186	23	Vacuum tubes	138	345	
TAT-5	1970	SF	One coaxial (1.5 in)	361	12	Transistors	845	2,112	
TAT-6	1976	SG	One coaxial (1.7 in)	694	6	Transistors	4,000	10,000	
TAT-7	1983	SG	One coaxial (1.7 in)	660	6	Transistors	4,200	10,500	
TAT-8	1988	SL280	Three-fiber pair	108	41	Lasers and repeaters	8,000	40,000	280 Mbps

Table 5.1 (continued)

System Name	Service Date	Cable	Repeaters	Two-way Voice Circuit Capacity	Digital Capacity
TAT-9	1992	SL560 Two-fiber pair	75 Lasers and repeaters		80,000 per pair	560 Mbps per pair
TAT-10	1992	SL560 Two-fiber pair	75 Lasers and repeaters		80,000 per pair	560 Mbps per pair
TAT-11	1993	SL560 Two-fiber pair	Lasers and repeaters		80,000 per pair	560 Mbps per pair
TAT-12/13 Ring	1996	SL2000 Two-fiber pair per span†	28 (45 km) Erbium-doped optical lamps	64,512 per span		5 Gbps per pair
TAT-14 Ring	2001 scheduled	Four-fiber pair per span†	45 km Wave-division multiplexing			160 Gbps per pair

* Retired from service.
† One pair is spare

History

The first transatlantic coaxial cable system, TAT-1, was installed by AT&T and went into service in 1958. It consisted of two coaxials, called SB type cable, one in each direction, and had a total system capacity of only 36 two-way voice circuits. A total of 51 electron-tube repeaters were used with this early system, and the underwater repeaters were spaced every 44 statute miles.

The next generation of submarine cable also utilized vacuum tubes but carried 138 two-way voice circuits on a single SD-type coaxial. This system was first used in 1963 on the TAT-3 transatlantic system between the United Kingdom and New Jersey, with repeaters spaced every 23 statute miles. A single SF-type coaxial and transistorized repeaters were used in 1970 on the TAT-5 transatlantic system, with repeaters spaced every 12 statute miles. TAT-5 was designed to carry 845 two-way voice circuits.

Time-assignment speech interpolation

The circuit capacity of these early, and later, submarine systems could be greatly increased through the use of a technique called time-assignment speech interpolation (TASI). TASI uses the silent intervals in speech conversations to carry signals from other speech conversations. This requires rather complicated and costly processing at each end of the submarine cable system, but the increase in circuit capacity is well worth the additional cost. As you can well imagine, it is quite costly and difficult to install submarine cable systems under oceans. With use of TASI, the early TAT-1 system could carry 72 two-way voice circuits, and the later TAT-5 system carries 2,112 two-way voice circuits.

Fiber systems

The transatlantic optical cable system TAT-8 was completed by AT&T in 1988 and utilized SL-type cable consisting of three pairs of single-mode optical fiber in a single cable, with one pair for protection. Each fiber pair carries 4,000 64-kbps two-way digital voice circuits. Digital processing that very efficiently capitalizes on the silent intervals in speech conversations—digital TASI—is used to increase the maximum system capacity to 40,000 two-way voice circuits. The repeaters are spaced every 41 statute miles, and 108 repeaters are used along the total length of the cable. The TAT-9 system, opened for service in 1991, has repeaters spaced ever 75 statute miles and carries a maximum of 80,000 two-way voice circuits.

The TAT-12/13 system

The newest undersea fiber for the Atlantic is the TAT-12/13 system, which became operational in 1996. The TAT-12/13 system was planned by a consortium of carriers consisting of AT&T, British Telecom, and France Telecom. TAT-12 and TAT-13 operate together as a system forming a loop with two spans under the Atlantic Ocean and two undersea spans at each end to complete the loop. The TAT-12 link connects Green Hill, Rhode Island with Lands End, England—a distance of 5,913 km. The TAT-13 link connects Shirley, New York with Penmarc'h, France—a distance of 6,321 km. Undersea local spans connect Rhode Island with New York and England with France to complete the loop. The under-Atlantic cable consists of two fiber pairs—one pair in service and the other pair as protection.

The under-Atlantic cables of the TAT-12/13 system carry 5 Gbps and utilize erbium-doped optical amplifiers. The trend has been to avoid TASI to assure the highest circuit quality. In the future, as demand increases, capacity on the TAT-12/13 system will be increased through the use of wavelength-division multiplexing, without the need to install any additional undersea cable. A threefold increase in capacity is expected.

No sooner is TAT-12/13 operational than the next system is being designed. TAT-14 will be a fiber ring with two spans under the Atlantic. Each span will contain four fiber pairs, with two pair as protection. The capacity of each pair is expected to be 160 Gbps.

Installation

Undersea cable is laid down by cable ships, such as the *C. S. Long Lines*, shown in Figure 5.15. The cable is usually placed in trenches by trenching machines when installed along the continental shelf to protect the cable from trawling or other damage. However, for other parts of the route, the cable is simply laid out from the ship and allowed to settle on the floor

Figure 5.15   Much of the optical fiber of the TAT-8 system was installed under the Atlantic by AT&T's cable ship *C. S. Long Lines*, shown here at dock. The ship was sold by AT&T in 1997 and is now owned by Tyco Submarine Systems, Ltd. (Photo courtesy of AT&T Bell Labs.)

of the ocean. Today's cable ships are kept quite busy laying cable under the Atlantic and the Pacific oceans and other seas. The large capacity and reliability of optical fiber make it the transmission medium of choice, and rates for international calls will continue to decrease as capacity continues to increase.

In 1997, AT&T sold its entire undersea cable operation to Tyco International. Tyco Submarine Systems, Ltd. is now responsible for the maintenance and installation of undersea cable systems. The actual systems are usually financed and owned by a consortium of international carriers.

Vocoders

There is considerable redundancy in a speech signal, and methods have been devised to reduce the analog bandwidth and digital data rate needed to transmit a speech signal. These reductions are performed by voice coders and decoders, called vocoders. Human speech is analyzed and converted to parameters that require less analog bandwidth or digital data rates for transmission. At the receiver, the received parameters are used to construct a synthetic speech signal. The use of vocoded speech is particularly essential in wireless communication, as previously mentioned in an earlier chapter.

The parameters used to encode the speech signal are derived according to which model is used to represent the speech. One model represents the speech signal according to its spectrum. A vocoder based on this spectral representation is called a channel vocoder. A representation of the source signal is also needed, along with a determination of the pitch of the voiced intervals of the speech signal. A channel vocoder can transmit a telephone-quality speech signal in 300 Hz, or in digital form, about 2,000 bps. This is a considerable savings over the 3,500 Hz required in analog form or the 64,000 bps in digital form.

Other models are used to represent the speech signal. The formant model represents the spectral information in terms of the resonant peaks, or formants, in the spectrum. The articulatory model represents the speech in terms of the positions of the tongue and lips. The popular linear-predictive-coding (LPC) model represents the speech signal in terms of the parameters of a digital filter. With the LPC technique, the speech signal can be compressed to 1,200 bps.

One problem with all these techniques is the cost and complexity of the coding and decoding equipment required for each voice circuit. The ever-increasing bandwidth of transmission systems simply has outpaced the vocoder technology. The major exception has been in wireless cellular systems, where the radio spectrum is very limited. The other application is the use of speech coders in voice encryption for security purposes. Perhaps the biggest problem with speech coding is that the synthesized speech is not natural sounding and frequently has a buzzy quality. A fair amount of research is being conducted to solve this problem, though.

Summary of transmission capacities

The various transmission media that have been used for long-distance telephone communication are as follows:

- Coaxial cable;

- Terrestrial microwave radio;

- Satellite microwave radio;

- Optical fiber and;

- Undersea cable.

The maximum route capacity of each of these media has grown over time at an increasing rate of growth, as shown in Figure 5.16, from the 1940s until the late 1980s. Today's optical fiber systems eclipse these capacities by large margins.

This is truly an exciting story of the ability of technology to offer continuous improvements resulting in ever-increasing route capacities. For the consumer, this has created a true bargain in long-distance communication.

The newest medium—and clearly already the winner—is optical fiber. The capacity of fiber continues to grow, and the distance between repeaters continues to increase. Transmission has become so cheap that domestic long-distance rates are distance independent. International rates will soon also be distance independent.

Transmission—an assessment

It is very clear that transmission has become very inexpensive, thanks to the vast capacity of optical fiber along with the ease of time-division multiplexing. AT&T no longer has a monopoly on long-distance service, and many carriers have entered the long-distance business all with their own networks. With all these entrants and all this capacity, a glut in circuits has occurred and long-distance rates will continue to decrease. Competition between the various carriers will become more intense, since in this world of digital one carrier's bit is the same as any other carrier's bit. Differences in rates, ancillary services such as operator service, and maintenance quality will be stressed to distinguish one carrier from another.

As rates continue to decrease, one wonders where it will all lead. Will residential traffic increase? Will people talk longer during social

Figure 5.16   The maximum route capacities of the various transmission media used in the AT&T long-distance network have increased greatly over time.

calls? Will facsimile calls substitute for mailed letters? How much time can people spend on the telephone?

Although technology has had great impact on long-distance transmission. the local loop remains old-fashioned, simple, twisted pairs of copper wire. The local loop with its restricted bandwidth and capacity is viewed as a communication bottleneck, but is it?

The key to communication capacity is understanding communication needs. Most communication services can be transmitted quite easily over the existing local loop. Voice communication is already adequately served, and, in fact, the residential local loop is idle most of the day. Most data services rarely require more than a few tens of thousands of bits per second, and this rate can be easily carried over the existing local loop. The only service that could tax the local loop is two-way video communication—the picturephone or videophone. However, the consumer need for the picturephone is highly suspect, and many people simply do not want to be seen while talking over the telephone. Some persons are so negative about being seen while speaking over the telephone that they would probably pay extra *not* to have a picturephone! More is said about these and other services in a later chapter.

Selected bibliography

Ching, Y., and H. S. Say, "SONET Implementation," *IEEE Communications Magazine*, Sept. 1993, pp. 34–40.

Desurvire, E., "Lightwave Communications: The Fifth Generation," *Scientific American*, Jan. 1992, pp. 114–121.

Flanigan, J. L., *Speech Analysis Synthesis and Perception (Second Edition)*, Berlin, Germany: Springer-Verlag, 1972.

Keiser, B. E., and E. Strange, *Digital Telephony and Network Integration*, New York, NY: Van Nostrand Reinhold Company, Inc., 1985.

Lee, B. G., M. Kang, and J. Lee, *Broadband Telecommunications Technology (Second Edition)*, Norwood, MA: Artech House, Inc., 1996.

Nagel, S. R., "Optical Fiber—The Expanding Medium," *IEEE Communications Magazine*, Vol. 25, No. 4, April 1987, pp. 33–43.

Omidyar, C. G., and A. Aldridge, "Introduction to SDH/SONET," *IEEE Communications Magazine*, Sept. 1993, pp. 30–33.

Romps, P., and A. Welles, "Unique Applications for VSATs," *Telecommunications*, Dec. 1991, pp. 25–26.

Seidman, L. P., "Satellites for Wideband Access," *IEEE Communications Magazine*, Oct. 1996, pp. 108–111.

Sipress, J. M., "Undersea Communications Technology," *AT&T Technical Journal*, Jan./Feb. 1995, pp. 4–7.

Trischitta, P., et al., "The TAT-12/13 Cable Network," *IEEE Communications Magazine*, Feb. 1996, pp. 24–28.

Willner, A. E., "Mining the optical bandwidth for a terabit per second," *IEEE Spectrum*, April 1997, pp. 32–41.

6

Electromechanical Switching Systems

Introduction

One telephone can reach any other telephone over a network that is switched. The switched nature of the telephone network—and data networks, too—is a large factor in its flexibility and utility. Switching is accomplished by switching machines, or systems, located throughout the network. These switching systems connect one telephone circuit to another.

This and the next chapter treat switching systems. This chapter begins with a general description of the evolution of network switching and how the network is organized from a switching perspective. Switching systems of the past accomplished switching by directly connecting one circuit to another through the use of electromechanical switches under the control of permanent wire instructions. Electromechanical

127

switching systems are no longer in use in the United States, and have been replaced by electronic switching systems, most of which accomplish switching of signals in the digital format. From a historical perspective, it is useful to understand the workings of the old electromechanical switching systems, and they are covered in this chapter. In addition, the chapter includes a discussion of how network traffic is measured and resources appropriately determined. Electronic switching systems are covered in the next chapter.

Evolution of network switching

In Bell's invention and first demonstration, telephone communication went from one instrument to another a few rooms away. This type of private-line service characterized initial telephone installations: one telephone could only communicate with one other telephone—not a particularly exciting form of communication. Early on, a number of telephones were connected to one line, a form of party line, so that a number of people could participate in the conversation, as shown in Figure 6.1. But there was no way to reach telephones connected to any other line. The solution was a means so that any telephone could be connected, or switched, to any other telephone.

One way, though impractical, of accomplishing such switching would be to bring pairs of wires, or lines, from all telephones to all other telephones. In this way, each telephone would simply connect by a simple switch to the line to the desired party's telephone, a form of station switching. This would be fine if only a small number of lines were involved, such as a simple intercom system for the home, but if hundreds of telephones needed to be reached, then hundreds of lines would need to go to each telephone, clearly an impractical arrangement.

The practical solution was discovered quite early and was the implementation of a centralized switching arrangement. All the lines from all the telephone stations were brought to a common place, a central office, where the electrical cross-connections could be made to connect one station to another. The central office was also called the telephone exchange. The actual connections were made by human operators who operated the exchange.

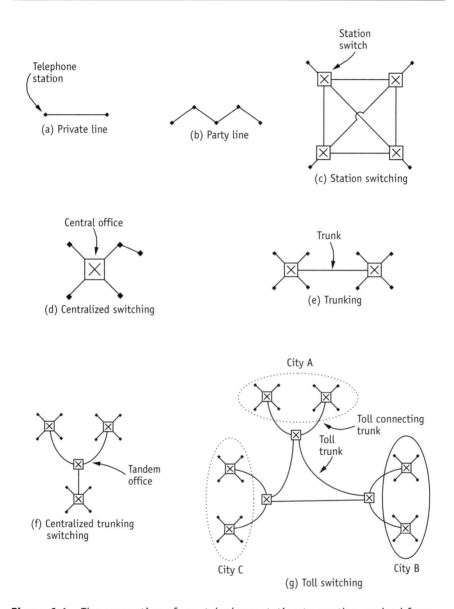

Figure 6.1   The connection of one telephone station to another evolved from simple, unswitched private-line service (a-b) to complex inter-city networks involving many stages of switching . Station switching (c) is impractical since lines from all other stations must be brought to each and every station. Instead, lines are brought to a central office (d), and calls involving a second office are completed over trunks (e). Tandem offices (f) serve a number of central offices, and toll offices (g) switch calls from city to city.

As exchanges grew to cover larger geographical areas, it became uneconomical to bring all the lines from outlying areas to a single central office. More central offices were created, each serving a nearby surrounding area. A call from a party served by one office to a party served by another office was completed over dedicated lines, called trunks, that interconnected the offices. The operator in one office reached an operator in the other office over a trunk circuit and relayed the identity of the called party so that the distant operator could complete the call. As growth continued, special switching offices, called tandem offices, were developed to handle the switching of interoffice calls involving trunks between local offices.

Extended switching systems were developed to serve long-distance and toll circuits between distant cities. Switching offices, called toll offices, were developed that performed switching solely for toll circuits.

For many businesses, most telephone traffic was between telephones that were all located on the customer's premises. This traffic could be served most efficiently by a private switching system located on the premises. This switch was called a private branch exchange, or a PBX. Today's PBXs are automatic (PABX), mostly using electronic technology (EPABX).

Telephone network switching

Switching offices

The telephone network is composed of a number of centralized locations, called switching offices, where the telephone conversation carried on one circuit is switched, or connected, to another circuit. The three major types of switching offices are local, tandem, and toll. The local office, sometimes called the end office or the serving office, is the one closest to the telephone station and connects directly to the local loop. A tandem office serves a cluster of local offices. Toll offices, or toll centers as they are sometimes called, are involved in the switching of traffic over long-distance, or toll, circuits.

Switching hierarchy

In the past, switching offices were organized into a hierarchy with the end offices where the call originates and terminates being at the lowest level.

The end office is classified as a class-5 office in the hierarchy. The end office connects to a toll center (a class-4 office); the toll center connects to a primary center (a class-3 office); a primary center connects to a sectional center (a class-2 office); and, finally, at the very top of the hierarchical pyramid, a sectional center connects to a regional center (class-1 office). This hierarchical structure is depicted in Figure 6.2. While there are well over 20,000 end offices, there were only about 20 regional centers serving the United States and Canada.

Normally, a call is completed using as few switching offices as possible. An end office will attempt to make the connection directly to the terminating end office over a direct interoffice trunk. If necessary, the switching capabilities of a tandem office will be used. If the call is a toll call, then the toll network and its toll-switching offices become involved.

With the hierarchical system, the preferred route was to connect across from an originating office to a lower terminating office using routes called high-usage routes. If this route was busy, then the next preferred route was to connect across to a center at the same level, or if this route too was busy, then across to the next higher center. As a last resort if all the routes across were busy, then the call would ascend to the next higher originating office and attempt to cross over again.

The routing used in today's long-distance networks is a dynamic nonhierarchical system in which all toll-switching centers are equal, and the route for each call is chosen according to the capacity that is available. Thus, with this newer system, the final routes are not fixed, but are chosen dynamically at the moment of the call. Increased flexibility and efficiency are possible with this dynamic nonhierarchical system.

Local office

The local central office is where the initial and terminating switching occur. As shown in Figure 6.3, all the local-loop wire pairs enter the central office at the cable vault, usually located in the basement of the building. The cables run in all four directions of the compass and usually enter the building from conduits. The cables then turn upwards through risers to upper floors of the building. A photograph of a cable vault is shown in Figure 6.4.

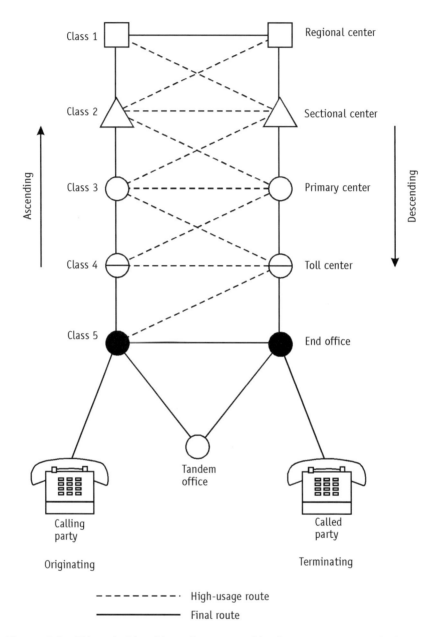

Figure 6.2 This switching hierarchy was used in the past. Present telephone network switching is a dynamic nonhierarchical system in which all the toll-switching centers are equal, and the route for each call is chosen according to what capacity is available.

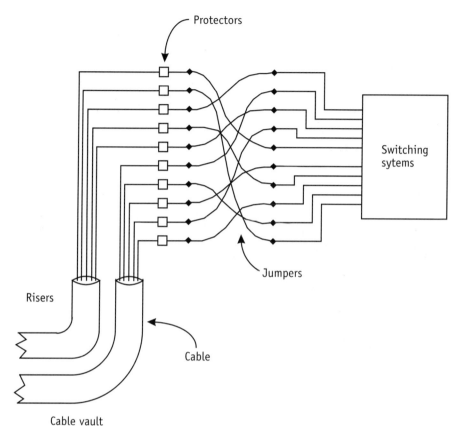

Figure 6.3 Jumpers connect from the protectors to the actual input to the switching equipment in a central office.

The sensitive switching equipment at the central office must be protected from possibly damaging overvoltages and currents, and hence each line has its own small circuit breaker and fuse, called a protector. All the protectors are situated in a protector frame, shown in Figure 6.5. Wires called jumpers interconnect from the protector frame to the actual inputs to the switching system. All these jumper wires are situated at the main distribution frame (MDF), shown in Figure 6.6. The jumble of long wires in older MDFs is minimized in the newer Lucent Technologies COSMIC® frame (for COmmon Systems Main InterConnect), shown in

Figure 6.4 Photograph of a cable vault. Cables enter the vault horizontally from underground conduits and then turn upwards toward the floors containing the switching equipment.

Figure 6.7. The interconnection of jumpers at the distribution frame is one of the last remaining manual tasks in a modern central office.

All the wires between the distribution frame and the switching system are usually placed neatly in troughs in the ceiling. All the wires, a pair for each access line served by the office, are still quite impressive. Subscriber loop carrier systems combine together a number of voice circuits in the field and bring them back to the office over a single pair of wires or over optical fiber.

A dial tone and the other tones used for signaling over the local loop are generated at a central place. The whole switching system and the local loop are powered by direct current (dc). In times of emergency, dc can be supplied by a large common battery located at the central office.

Figure 6.5 Photograph of a protector frame. Each line has its own protector to protect the switching equipment from excessive voltages and currents.

Normally, the alternating current (ac) supplied by the power company is converted to ac, or rectified, by equipment at the office. A diesel generator stands by to supply power to the office when the utility power has failed for more than a specified time and also to recharge the batteries. The large currents needed to power the system are carried over thick copper plates, called bus bars.

When I visit a central office, I am amazed in this day of modern electronics to still see such large amounts of copper wire. When I descend into the cable vault, I see that much of the cable is thick and black. However, I also see much thinner orange cable. This represents optical fiber that is being used for trunks to other offices and to interexchange carriers. Fiber is also being used in feeder systems in the local loop. When I visit the

Figure 6.6   Photograph of a main distribution frame (MDF). Here, jumper wires connect the protected lines to the switching machine.

Figure 6.7   Photograph of a Lucent Technologies COSMIC main frame. The jumble of wires of a conventional frame is avoided through careful design.

upper floors of the office, I am surprised at all the empty space. That space was formally occupied by old, electromechanical switching equipment, which was probably removed during the past 10 years or so. I miss seeing any old crossbar switching systems that give an intuitive feel for the mechanics of switching that it is impossible to obtain with today's electronic systems. Switching technology has greatly evolved over the years.

Technology evolution

During the first few decades of telephone service, switching was a manual operation performed by human beings, called operators, who made the actual connection of circuits. (See Figure 6.8.) The connections were made at a switchboard utilizing cords with plugs at the ends. The plug had a tip and a ring that completed the actual electric circuit over which the telephone signals traveled. The sleeve was used for signaling and supervisory purposes. The terms "tip" and "ring" continue to this day for the two wires between the central office and the telephone instrument.

Figure 6.8 With manual switching, a three-conductor connecting cord makes the electric connection from one telephone line to another. The cord terminates at each end in a plug in which the metallic transmission circuit is formed by the tip and ring of the plug. The sleeve is used to complete signaling circuits used within the central office for such purposes as indicating that a circuit is in use.

Each operator had about 18 cords that could be used to make connections. About 120 incoming lines terminated at the operator's switchboard at answering jacks. A lamp at each jack would light indicating that the line desired service. The operator would plug-in and answer the call. The calling party would give the name, and later the telephone number, of the called party. The operator would then complete the call by connecting the cord to one of perhaps as many as 10,000 subscriber multiple jacks within reach. Lamps associated with each connecting cord indicated when parties had completed a call.

Manual switching was slow and very costly in terms of its extensive use of human labor, as shown by the photograph of Figure 6.9. However, the use of humans gave it considerable intelligence and a friendly (we hope) interface with users.

The first major innovation in switching came in 1892 with the first installation of an automatic switch controlled by the subscriber's

Figure 6.9   Photograph of a manual switchboard capable of serving 5,000 lines. (Courtesy of AT&T Bell Labs.)

telephone instrument. The switch was conceived by Almon B. Strowger. A later modification of his invention included the invention of the rotary dial and the use of dial pulses to control the operation of the switch. The Strowger switch was an electromechanical device, and was later adapted for use in the Bell System, starting in 1919. Later, Bell System engineers were able to develop improved electromechanical switching systems. However, electromechanical switching systems were slow and noisy, and were not very flexible in terms of offering new services. But they most certainly were less costly than human operators, and were directly responsible for making telephone service affordable and universal.

Today's telephone switching systems are electronic switching systems using either analog space-division switching or digital time-division switching. The electronic technology makes faster connections, and the programmable computer that controls the switches offers considerable flexibility.

Switching system

There are two functional parts of any telecommunication switching system, as shown in Figure 6.10. The first part performs the actual switching and connection of one circuit to another. This part consists of the switches themselves, which are interconnected to create a network of

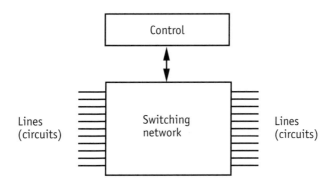

Figure 6.10   A telephone switching system consists of two major parts: the switching network itself, which is the actual means for connecting one circuit to another, and a means of control for instructing the switching network to make specific connections.

switches, commonly called the switching network. The second part of a telecommunication switching system is responsible for issuing the commands necessary to operate the switches at the proper time to make specific connections. This part is called the control function, or simply control.

Various switching systems have been designed and developed over the years. The switching network and the control in the earliest systems, such as the step-by-step system and the crossbar system, were completely electromechanical. Newer systems utilized electromechanical switching networks but electronic control in the form of a programmable digital computer; these systems were called electronic switching systems. The newest switching systems are totally electronic, having both an electronic switching network and electronic control.

We will examine the progression of switching technology through study of the workings of these various systems. But first we need to examine, from a theoretical perspective, the various ways in which switching is accomplished. Then we will describe specific switching systems that have been developed and used over the years, starting with the earliest electromechanical systems in this chapter and ending with today's digital systems in the next chapter on electronic switching systems.

Approaches to switching

There are two basic ways in which the switching performed by the network in a switching system can be accomplished: space-division switching and time-division switching. Space-division switching is described below; time-division switching is described in the next chapter on electronic switching systems.

In space-division switching, as Figure 6.11 shows, each telephone conversation has its own dedicated physical path through the switching network and the physical connection is maintained for the duration of the call. Switches that switch signals in actual space over separate electrical paths in the switching network are used in space-division switching systems. There are many physically separate electrical paths in a space-division switching system.

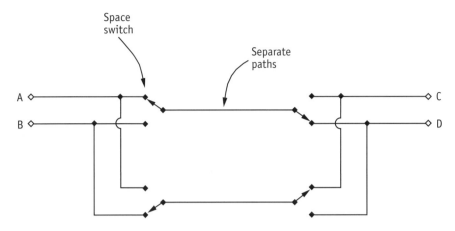

Figure 6.11 A simple space-division switching system utilizing two space switches to enable two circuits to be connected to two other circuits. Each electrical connection is established over separate physical paths through the switching network. In this example, party A is connected to party D, and B is connected to C.

Space-division switching is usually associated with the world of analog, and time-division switching with the world of digital. However, the switching universe is not quite that simple, and modern digital time-division switching systems usually utilize switches that switch signals both in space and in time. All this will become clearer in the next chapter when time-division switching is explained in detail.

Space-division switching

Basic switches

The electromechanical switch invented by Strowger was a stepped rotary switch that moved in two dimensions. It will be described in more detail in a following section on the step-by-step switching system. In its most general form, a rotary switch connects one line to any one of many lines, as shown in Figure 6.12. The switch rotates through its positions to make the final desired electrical connection. Though simple in concept, its fault

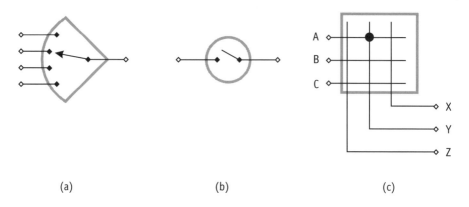

(a)   (b)   (c)

Figure 6.12   The basic types of switches used in space-division switching are (a) the rotary switch, (b) the on-off switch, and (c) the matrix switch.

is that electrical contacts must slide across each other, thereby subjecting them to considerable wear—and wear creates noise. Furthermore, the relatively large amount of mechanical motion means that a fair amount of time is needed to make an actual connection.

We are all familiar with the on-off switch that turns lamps on or off. In its electromechanical form, it consists only of two contacts and a means to complete or to open the circuit between them. Though simple, the electromechanical on-off switch is still subject to some wear, and though faster than a rotary switch, it is very slow in today's world of electronics. A transistor can be used as an on-off switch and offers the advantages of speed, low noise, and no wear.

On-off switches are used to create a basic coordinate, or matrix switch. A matrix switch can connect many lines with many other lines. Connections are made with on-off switches at the crosspoints in a matrix of all input lines and all output lines. The electrical connection at each crosspoint can be made using a variety of electromechanical and electronic technologies, such as conventional contacts, small reed contacts sealed in glass, and diodes and transistors.

Switching stages

Normally, not everyone will want to converse with everyone else at the same time. Hence, a switching network does not need to be designed to interconnect between all telephone lines simultaneously The incoming lines can therefore be concentrated and distributed over a smaller number of switching paths before being expanded at the last stages of switching, as shown in Figure 6.13. In this fashion, the switching is accomplished in stages consisting of concentration, distribution, and expansion.

A basic three-stage switching network using rotary switches is shown in Figure 6.14. The electromechanical Strowger rotary switch is able to connect one line to any one of 100 other lines and is the basic technology used in the switching network in the step-by-step switching system. The step-by-step system is explained in a following section.

A basic three-stage switching network using matrix switches is shown in Figure 6.15. The electromechanical crossbar switch is a matrix switch that can connect 10 inputs to 20 outputs. The crossbar switch, invented at Bell Labs, is used in the switching network in crossbar switching systems. The crossbar system is explained in a later section.

Many telephone calls are between lines served by the same switching office. To handle these calls, the output from the switching network is connected back on itself to the incoming lines to make these intraoffice connections. The lines used for such intraoffice calls are called intraoffice trunks.

Figure 6.13   There are three stages of switching in a space-division switching network. Many input lines are concentrated together to a smaller number of serving lines. The ratio of the number of input lines to serving lines is greater than 1. These serving lines are then distributed to output switches that expand service to a large number of output lines. The ratio of the number of output lines to input serving lines is greater than 1.

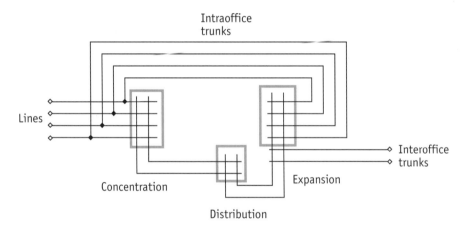

Figure 6.14   A basic rotary three-stage switching network. If more than two input lines demand service, they will be denied service since there are only two input switches available to serve them.

Figure 6.15   A basic matrix, or coordinate, three-stage switching network. Electric on-off switches make the required connections at the crosspoints. Calls within the office are made over intraoffice trunks that connect the output from the switching network back to the input lines.

Blocking

Clearly, it is possible that some common paths in the switching network might become completely congested so that no more calls can be carried.

In this case, any additional calls desiring service will be blocked. The blocking can occur at any stage, and the switching network must be designed to minimize blocking for the maximum traffic that it handles. The acceptable amount of blocking depends on the type of customer served by the switching office, insofar as different types of customers have different traffic patterns. In particular, business customers usually make many more calls during the day than residential customers.

It is possible to design a switching network so that it is nonblocking. However, such a design usually takes a large number of switches, and switches are costly. Through appropriate design, it is usually possible to implement a switching network with an acceptable amount of blocking and a fewer number of switches than for a nonblocking network. The example of Figure 6.16 shows a 12×12 matrix switch serving 12 input lines and 12 output lines. This matrix switch, requiring 144 on-off switches, is strictly nonblocking, and any input can be connected to any output. A three-stage switching network can be more efficient, particularly if some amount of blocking is acceptable. Only 84 on-off switches are needed in the three-stage blocking design. If rearranging the switch contacts during calls is acceptable, a nonblocking design requiring 120 on-off switches is possible.

Central office switching systems typically serve 10,000 lines. A strictly nonblocking matrix switch would require 100 million on-off switches and would be prohibitively costly. Since not all lines normally require service at the same time, more efficient multistage designs with acceptable amounts of blocking are used. Of course, there are those rare times when a high percentage of customers do demand service, such as in a disaster, and many customers will have to be turned away.

General types of switching systems

The two major technologies used in switching systems are electromechanical and electronic. Electromechanical is the way of the past; nearly all modern switching systems today are electronic.

Electromechanical switches are the foundation of space-division switching systems. The earliest type of electromechanical switching system was the step-by-step system using the Strowger rotary switch.

(a)

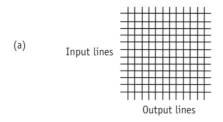

Input lines

Output lines

(b)

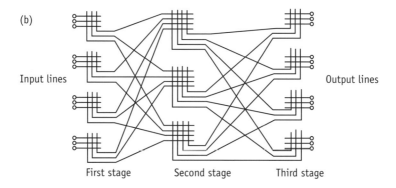

Input lines

Output lines

First stage   Second stage   Third stage

(c)

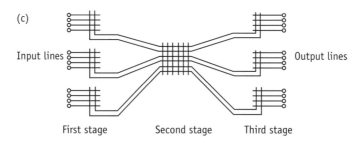

Input lines

Output lines

First stage   Second stage   Third stage

Figure 6.16   (a) A 12 × 12 matrix serves 12 input lines and 12 output lines requiring 144 on-off switches at the crosspoints. (b) Through efficient design, the number of switches can be reduced with a three-stage switching network to 120 on-off switches. (c) If some blocking is acceptable, the switching network can be even more efficiently designed, as in this example requiring only 84 on-off switches.

Another generation was the panel system, although it was relatively short-lived. The next generation was the crossbar system, which utilized a coordinate or matrix switch. Electromechanical control technology was used in the step-by-step, panel, and crossbar switching systems. This technology was permanently hard-wired, and features and functionality could not easily be changed. Electromechanical control was nonprogrammable and could not be reprogrammed.

The following sections describe the workings of electromechanical switching systems that were manufactured by AT&T's Western Electric Company and GTE's Automatic Electric Company for use in North America.

The step-by-step switching system

History and invention

The basic switch used in the step-by-step switching system was invented by Almon B. Strowger, an undertaker. The anecdotal story is that Strowger invented the switch since he was upset that all the business in his town went to his competitor, whose wife was the operator of the local exchange. His invention was an automatic switch that stepped vertically and horizontally under the control of digits entered at the telephone instrument. His switching system was first installed in La Porte, Indiana in 1892. A few years later, in 1896, the telephone dial was invented by some of Strowger's associates.

The Strowger switching system was manufactured by the Automatic Electric Company, formed in 1901, and was sold primarily to the non-Bell, independent telephone companies. The first installation of a Strowger switching system in the Bell system did not occur until 1919. Until then, the Bell companies had continued to use human operators, and had ignored the new technology for nearly 25 years. The Strowger system was called step-by-step in the Bell System.

Until the late 1980s, it was still possible to find step-by-step switching systems in use in the Bell System and in independent companies. However, step-by-step switching is extinct today in the United States. Step-by-step switching had a lifetime of nearly 100 years, which clearly indicates the reliability and utility of Strowger's invention. It is very rare indeed to find such a long lifetime for an electromechanical technology.

The Strowger switch

The key ingredient in the step-by-step switching system is the basic Strowger switch, shown in Figure 6.17. The Strowger switch is an ingenious, electromechanical device formed from electromagnets and ratchets. The electromagnets are activated by the dial pulses, and ratchets cause a wiper assembly to step vertically and then to rotate horizontally in steps until the desired contact is reached (see Figure 6.18). There are 100 contacts arranged in a bank of 10 rows by 10 columns.

There are two banks of contacts in a Strowger switch. The bottom bank is called the line bank, and it is where the actual tip and ring connection is made for the telephone conversation. The top bank is called the sleeve bank, and it is where the control connection is made. A direct

Figure 6.17 A photograph of a basic Strowger switch. The contact banks and wiper arm are shown along with the relays at the top that control the movement of the wiper arm.

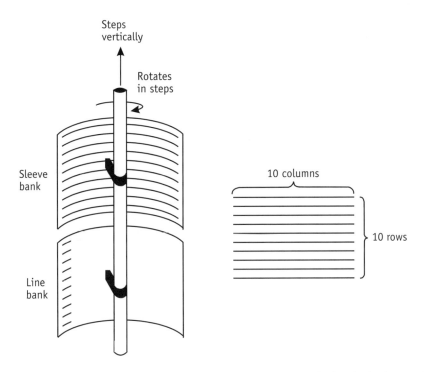

Figure 6.18 A wiper arm moves vertically in steps to reach the desired row of contacts. It then rotates horizontally in steps until the exact column is reached, thereby making the desired connection.

current flows through the sleeve contacts to maintain an actual connection. When the calling party hangs up, the flow of current through the sleeve circuit ceases and a spring causes the switch to return to its resting position.

There are a number of relays at the very top of a Strowger switch. These relays count the dial pulses and control the electromagnets that cause the wiper to move.

The action of the wiper as it slides across the contacts results in considerable wear of the contacts, which introduces a fair amount of electrical noise into the connection. There are many mechanical adjustments that need to be made to keep the switch working properly. Maintenance is a nearly continual process. Nevertheless, the basic simplicity and high reliability of the Strowger switch led to nearly a century of use.

Step-by-step switching system

A step-by-step switching system usually serves 10,000 telephone lines. For an office serving 10,000 lines, a typical step-by-step switching system is a very large machine requiring many bays of equipment, as shown in Figures 6.19 and 6.20.

Figure 6.19   A step-by-step switching system is composed of many frames of Strowger switches. The clicking noises of these switches can reach a crescendo during busy periods.

Figure 6.20 A considerable amount of wire interconnects the rear of the Strowger switches in a step-by-step switching system.

We next describe the operation of a step-by-step switching system that can reach any one of 10,000 lines, numbered 0000 through 9999, by the dialing of a four-digit telephone number. The dialed number corresponds exactly to the location of the contacts of the called party's line in the switching system.

The switching network is divided into stages, illustrated in Figure 6.21. The first stage is the line finder. When the calling party lifts the telephone handset from its cradle, the telephone instrument starts to draw direct current from the telephone line. The initiation of this flow of dc causes relays in the switching system to initiate a hunt by the line finder to locate the particular line that is drawing current and thus wishes service. Each line-finder switch serves 100 lines. Once a particular line-finder switch finds a line and connects to it, that line-finder switch is engaged and can no longer serve any other lines. Therefore, a number of line-finder switches are connected in parallel to serve the same 100 lines. The specific number of line-finder switches depends on the anticipated traffic and usually is 10 or less per 100 lines.

Each line-finder switch is hard-wired to the next switch in the sequence of switching stages. This next stage is called the first selector. The first selector connects the calling party to dial tone and then waits for dialing to commence. The first dialed digit causes the first-selector switch to move vertically a number of steps corresponding to the dialed digit. The switch senses the pause between dialed digits and then searches, or hunts, horizontally for an idle line to a second selector switch.

The second selector switch moves vertically with the second dialed digit and then hunts horizontally for an idle line to the last switch in the

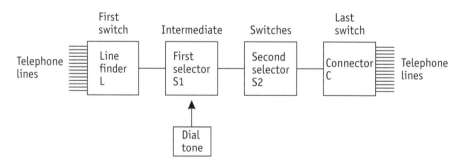

Figure 6.21 The switching network in a step-by-step system designed to reach any one of 10,000 telephone lines is accomplished in stages. The line finder searches for the direct current flowing in the line desiring service. The first selector returns dial tone and responds to the first digit. The second selector responds to the second digit. The last two digits operate the connector, which makes the final connection to the dialed party's line.

process. If either selector switch cannot find an idle line, the wiper passes by the tenth column of contacts and reaches an eleventh contact, which returns a fast busy tone to the calling party.

The last switch is called a connector. It first moves vertically with the third dialed digit and then rotates horizontally with the fourth dialed digit. The connector switch makes the final connection to the number dialed.

The above-described switching system can connect to any one of 10,000 lines in a single switching office. Other offices in a local exchange area are assigned three digits to indicate the specific office and switching system. These three office digits are then followed by the four-digit number of the line within the office. The calling party thus dials seven digits. Three additional selector switches are added to the step-by-step system to accommodate the office digits. If the call were to another office, the third selector switch would connect to an interoffice trunk to that office, and the last four dialed digits would make the final connection at the distant office.

Control

The control function in a step-by-step switching system is distributed throughout the system in the form of the relays that control each individual Strowger switch, as shown in Figure 6.22. There is no central control in a step-by-step system; each switch has its own control associated with and incorporated into it. The dial-pulse digits contain the information necessary to operate each switch. These dial pulses are carried over the same lines that will carry the speech signal. Thus, the control path is the same as the speech path.

The earliest step-by-step systems used a method of control called direct progressive control. With this method of control, each stage of switching responds directly to the digits dialed by the calling party. As one stage completes its connection, the next stage responds progressively to the next dialed digit. The connection is completed progressively through the switching network.

Direct progressive control is simple, but it has two problems. Because the paths through the switching network are created progressively, there is no way to look ahead to determine whether future paths

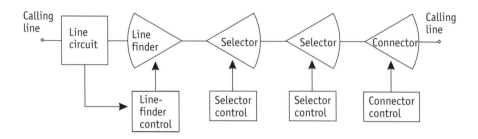

Figure 6.22   The control in a step-by-step switching system is distributed throughout the system in the form of the relays that control each individual Strowger switch. Each dialed digit directly controls the movement of each Strowger switch as the path through the switching network is progressively made.

are already in use. Thus, a call might be blocked since no idle paths are available. This means that additional spare capacity is needed to minimize the effects of blocking. Because the dialed digits control the switches directly, there must be a strict correspondence between the dialed digits, which represent the telephone number of the called party, and the physical contacts to which the called line is connected. The physical number is called the office number, and the listed number is called the directory number.

The two problems of direct progressive control are avoided with register progressive control. As shown in Figure 6.23, the line-finder switch connects the calling party to a register to receive and temporarily store all the dialed digits. After the called number has been completely dialed into the register, the register is connected to the selector and, in effect, "redials" the number. If blocking occurs, the number is still available in the register so that another redialing attempt can be made. A translator can be used with the register to translate the directory number into the office number needed to control the switches to reach the appropriate contacts. This means that the actual physical placement of a telephone line on the switch contacts does not need to be changed if a telephone number is changed, and vice versa.

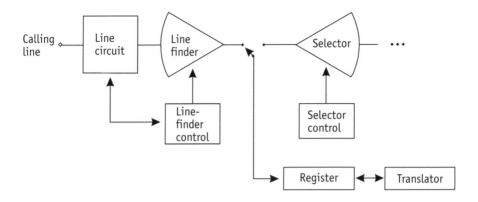

Figure 6.23 With register progressive control, the dialed digits representing the directory telephone number are entered and stored temporarily in a register. The register-stored digits can then be translated to the office number, which represents the actual location of the called party's contacts in the switching system. If blocking occurs, the register-stored digits can be reused for a second attempt.

The crossbar switching system

History and invention

The Bell System's answer to the Strowger system was the panel switching system, which was first installed in 1921. The basic switch in the panel system was a Byzantine affair able to access any one of 500 contacts, compared to Strowger's 100. This was accomplished by rotating cork rollers that moved the selecting contacts horizontally and vertically, much like the motion of a human operator at a switchboard. The basic switch created considerable electrical noise and need much maintenance. Although thousands of panel systems were installed in the United States, because of the maintenance costs to keep it operating, it was a failure in its day.

The real innovation in electromechanical switching was the Bell System's crossbar switching system, first installed in 1938 as the No. 1 crossbar system for large metropolitan offices. The No. 5 crossbar system, intended for rural and suburban offices, was first installed in 1948. A rural-office crossbar system, the No. 3 crossbar system, was first installed in 1974. Crossbar switching has had a long life and was still in use in the early 1990s.

The crossbar switch

The basic crossbar switch is a matrix, or coordinate, switch with 10 horizontal rows and 20 vertical columns, as shown in Figure 6.24. Input lines appear on the 20 verticals, and the output connections are made on the 10 horizontals. Any vertical can be connected to any horizontal, and 10 simultaneous connections can be maintained by the switch. The actual electrical connections between verticals and horizontals are made by electromechanical contacts at the crosspoints that open and close.

The basic operation of the switch is ingeniously simple, yet is difficult to visualize since the essential movements are very small and are hidden within the switch itself. As shown in Figure 6.25, there are five horizontal selecting bars that can each be rotated, or tipped, upwards or downwards through a small angle. These selecting bars are operated by electromagnets that receive their signals from the control section of the switching system. The verticals are addressed by 20 vertical hold bars, likewise operated by electromagnets that receive their signals from the control section. Operating the appropriate selecting bar either upwards or downwards and engaging the appropriate hold bar causes an electrical cross-connection to be made at the corresponding vertical and horizontal position.

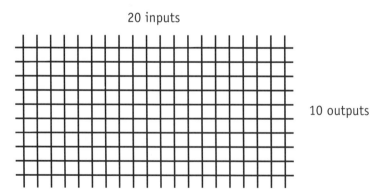

Figure 6.24   The basic switch used in the crossbar switching system is a matrix switch with 20 inputs along the verticals and 10 outputs along the horizontals.

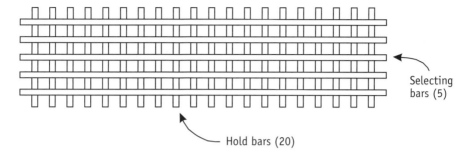

Figure 6.25   Five horizontal selecting bars each can rotate upwards or downwards through a small angle in the crossbar switch. Twenty vertical hold bars can each be engaged to make a specific connection at the desired cross-connect.

Twenty short pins of flexible wire, called fingers, protrude from the back of each selecting bar (see Figure 6.26). These fingers protrude backwards into the switch assembly. As the selecting bar is tipped upwards or downwards, these fingers likewise move up or down. The fingers thus have three possible positions: at rest, up, or down. The hold bars rotate and press upon the fingers, causing them to move sidewards.

The switch contacts at each crosspoint are attached to a piece of phenolic material that has a small notch at the end close to the finger. The contacts are in pairs. The phenolic is notched so that if the finger is in the rest (or unselected) position, the hold bar will push the finger sideways between both phenolic pieces without touching either of them. However, if the finger is moved up or down by rotation of the selecting bar, the finger will engage the notch and will be held there by the action of the hold bar. This causes the switch contacts to close. The selecting bar can be released, but the hold bar will lock the flexible finger in position, thereby maintaining the connection. Other connections can then be made with the same selecting bar by activating other hold bars.

The contacts in a crossbar switch need only to close and open, which requires very little physical movement. Wear and associated noise hence is at a minimum. Since more than one connection can be made by each crossbar switch, fewer switches are needed to make a switching network. The very simplicity of the crossbar switch minimizes costly maintenance. These advantages coupled with the use of a sophisticated control

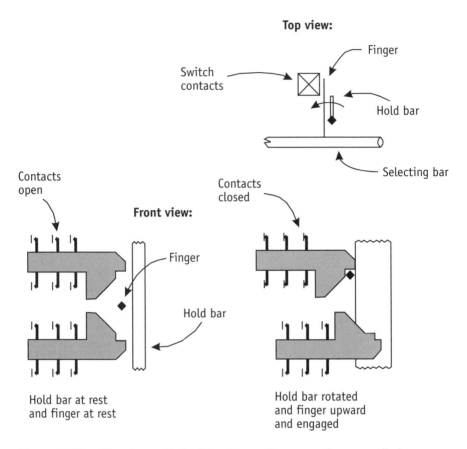

Figure 6.26   Twenty small, flexible pins, or fingers as they are called, protrude along the back of each selecting bar. When the chosen hold bar rotates, it clamps onto the appropriate single finger and locks it in place in the switch contact assembly. The engaged finger slips into a notch in a piece of phenolic material. The sideways movement of the finger causes the phenolic to move, and this causes the switch contacts attatched to the phenolic to be closed.

mechanism made the crossbar switching system a true winner. Finally, the Bell System had its answer to the Strowger system.

Crossbar system

The crossbar switching system uses a centrally located control mechanism that is shared in common by all the switches in the switching network. This type of control, therefore, is called common control. The actual control is performed by electromechanical relays called markers. The markers perform translation, test switches to determine whether they are in use, test different paths through the switching network, and issue the signals necessary to operate the switches. In effect, the markers are permanently programmed (called hard-wired) computers based on relay technology. The common-control markers are used only briefly on each call. Once used, the markers drop out, and the switching network maintains the voice connection for the duration of the call.

As shown in Figure 6.27, the switching network is organized into two distinct groups of multistage crossbar switching matrices. The line links connect to the subscriber lines, and the trunk links connect to intraoffice

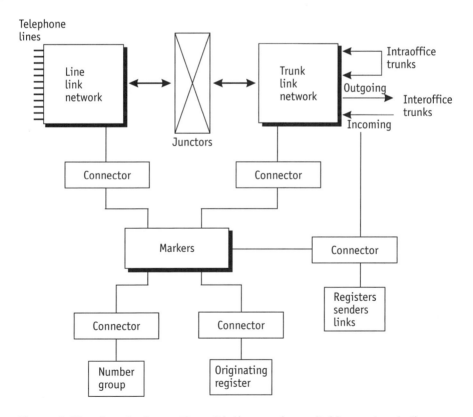

Figure 6.27 A major innovation with the crossbar switching system is the use of sophisticated control that is shared in common by all the switches and calls.

and interoffice trunks. The line links and trunk links are situated in frames (see Figure 6.28), and the two frames are connected together by wires called junctors. The junctors are situated together in their own frame.

The markers that control the line links and the trunk links access them through connectors. The originating register supplies a dial tone and stores the dialed telephone number. Telephone numbers are translated into locations on the line-link frame by the number group. An intraoffice call is connected back to lines within the office by intraoffice trunks at the trunk-link frame. Interoffice trunks are controlled by the markers through connectors and various registers.

Figure 6.28   Photograph of a crossbar switching system. (Courtesy of AT&T Bell Labs.)

Traffic

Switching and transmission systems are designed to handle the expected traffic. If a particular system is designed to be nonblocking, then all calls will be served. Most switching systems block some calls if they handle a large traffic volume.

Some useful measures of traffic include the total number of calls served during a specified time period, usually the busiest hour for the system, called the busy hour. Usage can be specified in terms of the percentage of time a trunk or switch is in use. Overflow is measured as the percentage of calls that found some particular piece of equipment or the whole system busy.

Telephone traffic is measured in units of hundred call seconds, or CCS. The product of the number of calls times their average duration in seconds gives the traffic in call seconds. Dividing this number by 100 gives the traffic in hundred call seconds. For example, 20 calls each with an average duration of 120 seconds would generate $(20 \times 120)/100 = 24$ CCS. The traffic density is the number of CCS per time period, usually the busy hour. For example, a PBX might be specified as being able to handle a traffic density of 1,500 CCS per busy hour.

Another measure of telephone traffic is the erlang, named after the Danish engineer and mathematician, A. K. Erlang. One erlang is equivalent to one call lasting one full hour. Because there are 3,600 seconds in an hour, one erlang is the same as 36 CCS. If there were 100 percent occupancy of a telecommunication path or device, then the traffic carried would be specified as either 1 erlang or 36 CCS. The PBX with a busy-hour capacity of 1,500 CCS could equivalently be specified as having a busy-hour capacity of $1,500/36 = 42$ erlangs.

Traffic density is usually associated with a specific grade of service. The grade of service is the probability that a new call will encounter blocking. For example, a grade of service of 0.01 for a traffic density of 1,500 CCS means that, at a traffic density of 1,500 CCS, one out of every 100 calls will be blocked.

A useful measure is the load factor, which specifies the average number of simultaneous calls during a specific hourly period, usually the busy hour. The load factor, C, can be calculated as the product of the number of calls originated, n, multiplied by the average holding time in seconds, h, divided by 3,600, or

$$C = nh/3,600$$

If the traffic is given in CCS, then the load factor is the traffic divided by 36. Consider a PBX that can carry 1,500 CCS during the busy hour with a 0.01 grade of service. The load factor is $1,500/36 = 42$. This means that the PBX can carry 42 simultaneous calls in the busy hour with a probability of blocking of 0.01.

Clearly, telecommunication systems can become congested. Models based on probability theory have been developed to characterize such congestion. The models relate the probability of blocking to the volume of traffic for different numbers of switching links, trunks, or servers, as they are called.

Two popular models for engineering switching systems are the Poisson model and the Erlang-B model. The Poisson model assumes that all blocked calls are retried within a relatively short period. This model is appropriate for describing the call attempts by a large number of customers, such as those served by a central office switching system. The Erlang-B model assumes that blocked calls are abandoned; it is appropriate for describing interoffice trunks. Tables are available that tabulate the capacity for different grades of service and different numbers of servers.

An example might help clarify the use of these models. Consider a PBX for which the total traffic during the busy hour has been measured to be 50 calls with an average duration of 5 minutes per call. The total busy-hour traffic is $(50 \times 5 \times 60)/100 = 150$ CCS. The acceptable grade of service is 0.01, which means that 1 call out of 100 will be blocked. The traffic and grade of service are looked up in the Poisson table, which shows that 10 servers are sufficient to handle the traffic. This means that a PBX capable of handling 10 simultaneous calls is sufficient, according to the Poisson model.

Selected bibliography

Ash, G. R., "Dynamic Network Evolution, with Examples from AT&T's Evolving Dynamic Network," *IEEE Communications Magazine*, July 1995, pp. 26–39.

Gurrie, M. L., and P. J. O'Connor, *Voice/Data Telecommunications Systems*, Englewood Cliffs, NJ: Prentice-Hall, Inc., 1986.

Schindler, G. E., Jr. (ed.), *A History of Engineering and Science in the Bell System: Switching Technology (1925-1975)*, Murray Hill, NJ: Bell Telephone Laboratories, Inc., 1982.

Talley, D., *Basic Telephone Switching Systems*, Rochelle Park, NJ: Hayden, 1979.

7

Electronic Switching Systems

Introduction

In the most general sense, electronic switching refers to the use of a programmable electronic digital computer to control the operation of the switching system, where the switching fabric could be space division or time division. Time-division switching is also known as digital switching. Electronic switching systems are covered in this chapter, starting with a general description of time-division switching.

Time-division switching

Basic principles

In time-division switching, as shown in Figure 7.1, separate conversations share the same electrical path through the switching system but are separated in time. In essence, the path is time-shared. Each conversation

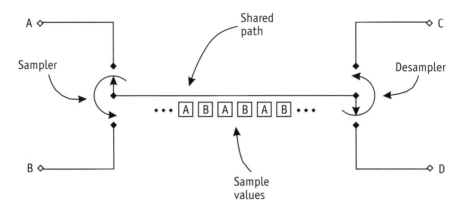

Figure 7.1 In time-division switching, a path is shared by many circuits for very short time intervals. In this example, the signal on circuit A is sampled by a sampler switch, and these samples values are interspersed between sample values from circuit B for transmission over a common, shared path. At the other end, the sample values are momentarily connected to circuits C and D in an order such that samples from circuit A go to circuit D and samples from circuit B go to circuit C.

is sampled, and the sample values for each conversation are packed together over the shared path for transmission. At the destination, the sample values are separated and routed to the appropriate conversation. Clearly, time-division switching is most appropriate for signals that have been sampled and digitized. For this reason, time-division switching systems are commonly called digital switches.

With space-division switching, signal paths are switched in physical space. With time-division switching, sample values of a number of signals sharing a common path are reorganized, or switched, in their time sequence.

Quite clearly space-division switching is somewhat akin to space-division multiplexing, and likewise time-division switching is akin to time-division multiplexing. One might therefore wonder whether there is some form of switching that is akin to frequency-division multiplexing, namely, frequency-division switching. With frequency-division switching, different frequency channels are assigned to different connections. This is the way cellular telephony works, as is explained in a later chapter.

Each telephone conversation uses a specific frequency channel in the radio spectrum. This type of switching is also particularly applicable to a single broadband medium shared by a number of users, such as the coaxial cable of CATV.

Time-slot interchange

In the chapters on transmission, we studied how a number of voice signals can be multiplexed onto a single transmission path using time-division multiplexing. Each voice signal is converted to a digital format, and each digital circuit is then assigned its own unique time slot in the stream of digital information. To switch these circuits, it is only necessary to interchange the digital signals in the time slots.

Consider a transmission path carrying the signals from four digital voice circuits. The digitized sample values, A1, A2, A3, and A4, corresponding to the four digital circuits flow sequentially along the transmission path in the order A1-A2-A3-A4, and so forth, as shown in Figure 7.2. The eight bits of digital information, A1, in the first time slot correspond to the first circuit, the second time slot corresponds to the second circuit, and so on. This stream of digital information forms the input to a time-division switch. The output from the switch is likewise a stream of digital information. However, the sequence has been reordered by the switch. The first time slot now contains A2, the second time slot contains A4, the third time slot contains A3, and the fourth time slot contains A1.

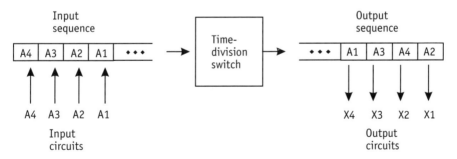

Figure 7.2 Time-division switching is accomplished simply by reordering the sequence of information contained in the time slots in a digital transmission path. In this example, input circuit A1 is connected to output circuit X4; A2 to X1; A3 to X3; and A4 to X2. This technique is called time-slot interchange.

The information in the first time slot in the output sequence is connected to the first output circuit, X1; the second time slot is connected to the second output circuit, X2; and so forth. Thus, in this example, since the first time slot in the output sequence contains A2, the second input circuit has in effect been connected to the first output circuit, X1. Similarly, A4 has been connected to X2, A3 to X3, and A1 to X4.

In this instance, switching has been accomplished through the simple interchange of information contained in time slots. This is called time-division switching, and it is performed by a technique called time-slot interchange. Since the signals switched in a time-division switching system are digital, time-division switching systems are sometimes called digital switches.

Time-slot interchange is performed through the use of a temporary storage called a buffer memory. The information contained in the input time slots is entered in sequence into the buffer memory, as shown in Figure 7.3. The information is then read from the buffer memory in a different sequence corresponding to the desired order of switching.

It takes time to write information into memory and then read the information from memory. This amount of time is called the write/read memory cycle time, and it determines the maximum number of time

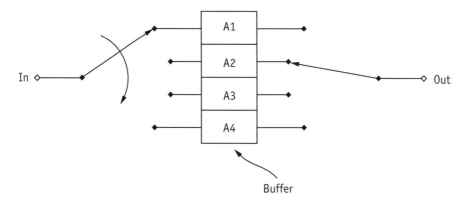

Figure 7.3   Time-slot interchange is performed through the use of a buffer memory into which the input samples are entered in the sequence received. The samples are then read from memory in a different sequence by a programmable switch. The order of the output switch in this example is 2-4-3-1, thereby reordering the output samples to correspond to the switching of Figure 7.2.

slots that can be switched in the buffer. The information in a particular location in the buffer usually waits awhile before being read out. This waiting time contributes a delay to the signal. The maximum delay is the total length in time of all the time slots stored in the buffer. Since usually the time slots for a full frame is stored in the buffer, the maximum delay is the length of a frame, which is 1/8,000 sec, or 125 μs.

Normally, a large number of digital circuits are involved in a telecommunication system. These digital circuits come from digital transmission trunks onto which a number of digital circuits have been multiplexed through time-division multiplexing. Or, the digital circuits can come from a number of analog voice circuits that have been digitized and multiplexed onto a digital line solely for the purpose of being switched in a time-division switching system. Either way, usually a number of physically separate multiplex lines are involved. Because a circuit on one multiplex line might need to be switched to a circuit on another multiplex line, some form of physical switching between different digital multiplex lines is needed. Such physical switching is accomplished using a space switch, and hence time-division switching systems usually involve a combination of both space switches and time switches.

The space switch used in a time-division switching system is usually a matrix switch that must continuously alter its specific connections according to which time-slots from different physical lines must be switched to other lines. Because this space switch is being continuously switched in time and shared by many lines, it is truly a time-multiplexed space switch and is called a multiplexed space switch. The following example explains the operation of a digital switching system operating in three stages: time, space, and time (TST).

Example of a digital TST switching system

The switching networks used in most digital switching systems use both time-division switching and space-division switching. The sequence of time switches (T) and space switches (S) categorize the switching system. The following example has three stages of switching in the following order: time-division switching (T), space-division switching (S), and time-division switching (T). This type of switching system is called a TST switching network

An example, adapted from Hills (1978), should clarify these concepts. Refer to Figure 7.4. Two digital trunks, A and B, each carrying four digital circuits are to be connected to two other digital trunks, X and Y, each likewise carrying four digital circuits. The connections are to be made as follows:

A1↔X4   B1↔X1
A2↔Y3   B2↔Y1
A3↔Y4   B3↔Y2
A4↔X2   B4↔X3

The input digital trunk lines A and B are each connected to time-slot interchange units. The digital samples from each line are entered into the two buffers in the sequence received. The output digital trunk lines X and Y are likewise connected to two time-slot interchange units, and the samples for each output line are read from the buffers in sequence from the first location in memory to the last.

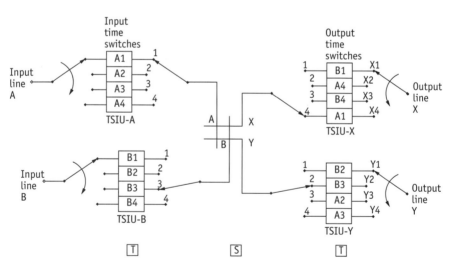

Figure 7.4 This digital switching network connects the four circuits in each of two input lines to four circuits in each of two output lines. The switching is accomplished in three stages: time-division switching, space-division switching, and time-division switching. (*Source:* M. T. Hills, *Telecommunications Switching Systems*, MIT Press, 1978, p.289, Figure 12.1.)

A multiplexed space switch with two inputs and two outputs connects the four time-slot interchange units. The space switch continuously reconfigures itself to send the appropriate sample values in the input buffers to the appropriate locations in the output buffers. The input data is arriving at a rate of four time slots every 1/8,000th second, or 32,000 time slots per second. Thus, the multiplexed space switch needs to reconfigure itself 32,000 times per second. The switch time is the time between each reconfiguring, and in this example the switch time is 1/32,000 sec, or 31.25 μs.

The information stored in the input time-slot interchange units (TSIUs) is read from the buffers in an order keyed to the switch order of the space switch. Similarly, the information entered into the output TSIUs is keyed to the operation of the space switch. The whole switching network must perform in perfect synchrony according to a master switching plan. The switching plan for the simple TST system is shown in Table 7.1. For example, at switch time 2, the information contained in time slot A4 of the TSIU serving line A is switched by the space switch to TSIU serving output line X. That TSIU stores the information in the second slot for later connection to X2. Thus, A4 has been switched to X2.

Clearly, even for such a simple example, this is a complicated process, requiring precise switching order of all the switches in the network. This switching order is determined by the computer controlling the switching system in response to the switching needs of the various circuits. Since all input circuits can be connected simultaneously to all

Table 7.1
The Various Time and Space Switches in the Switching Network of Figure 7.4

Switch Time	Space Switch		Time Switch			
	A	B	A	B	X	Y
1	X	Y	1	3	4	2
2	X	Y	4	2	2	1
3	Y	X	3	4	3	4
4	Y	X	2	1	1	3

output circuits, the TST switching network in this example is a nonblocking switch.

Other approaches

Large buffers were costly in the past, and also the write/read cycle time was somewhat long. Thus, large buffers were avoided through the use of multiplexed space switches to connect digital circuits on different input and output lines. However, storage continues to decline in cost, and write/read cycle times likewise continue to decline. Therefore, large buffers and fewer time-multiplexed space switches will be used in future digital switching systems.

The preceding example used a combination of time and space switching to accomplish the switching. The same switching could be accomplished through the use of two large buffers, as shown in Figure 7.5(a). The input samples from the two input trunks are entered in parallel into both buffers. Each buffer then serves one output trunk. An alternative approach, assuming a buffer memory that could have multiple simultaneous inputs and outputs, would be the use of a single buffer memory, as shown in part of Figure 7.5(b).

Specific switching systems

The solution to the inflexibility of electromechanical control was the use of programmable, electronic, digital computer technology for control of the switching system. Switching systems using such control were called electronic switching systems (ESSs) in the Bell System. The Bell System's No. 1 ESS™ system used electronic control, although the switching network was a space-division system initially using electromechanical switch contacts. Later, electronic diodes were used in the switch matrices.

Modern switching systems use time-division switching for the switching network along with computer control. The control can be located centrally, as in the Lucent Technologies No. 4 ESS™ system, or distributed throughout the switching system, as in the Lucent No. 5 ESS™ switching system. The DMS™ family of digital switching systems manufactured by Nortel (formerly called Northern Telecom) use a combination of distributed and central control.

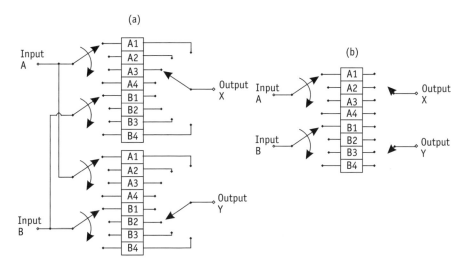

Figure 7.5   The time-multiplexed space switch can be avoided through the use of larger buffer memories: (a) two buffers are used for the two output lines, and all the input time slots are entered in parallel into both; (b) a single buffer with dual-access inputs and outputs accomplishes the same switching.

These specific systems are explained in the following sections of this chapter. Other vendors manufacture digital switching systems, but the Lucent Technologies and Nortel machines are most prevalent in North America, and hence these machines are featured.

Specific electronic switching systems

AT&T No. 1 ESS™ switching system

The markers in a crossbar system are hard-wired computers, and their nonprogrammable inflexibility is a serious disadvantage. The idea of using a digital computer, with its inherent programmable flexibility, to control the operation of the switching network was a natural progression in the evolution of switching technology. Digital computers are electronic devices, unlike the electromechanical relays used in markers in the crossbar system, and hence the new generation of switching technology was called an electronic switching system (ESS).

The first electronic switching system developed by AT&T was the No. 1 ESS system. It was first installed in 1965, and was designed to serve from 10,000 to 65,000 lines at a maximum of 25,000 calls in the busy hour. The No. 1A ESS system was introduced in 1976, and served as many as 128,000 lines at a maximum of 240,000 calls in the busy hour.

Reed switch

The speed and flexibility of a programmable computer to control the switching network necessitated a new basic switch. The new switch was a reed switch, consisting of two reed contacts sealed in a small glass envelope, as shown in Figure 7.6. The application of an external magnetic field causes the contacts to close and open. The reed switches are organized into basic 8 × 8 coordinate arrays, or switching matrices.

Two approaches have been used to operate the reed switches. The initial approach uses two plates of a special two-state magnetic alloy, called Remendur, which are mounted on both sides of the reed switch. The special alloy has the property that when momentarily magnetized in either polarity by the application of an electric magnetic field, the alloy retains the magnetism and polarity. In effect, the alloy is a magnetic memory. In one polarity, the alloy causes the reed contacts to close; in the opposite polarity, the alloy causes the reed contacts to open. Short pulses of electric current change the magnetism of the alloy. The assembly of reed switch and Remendur alloy is called a ferreed switch.

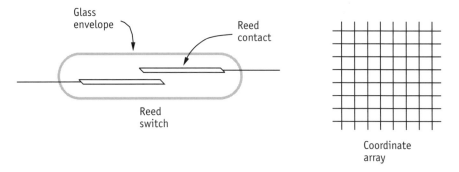

Figure 7.6   A simple reed switch is used in the AT&T No. 1 ESS switching system. The reed switches are organized into basic 8 × 8 coordinate arrays, or switching matrices.

A later approach used the Remendur alloy for the reeds themselves. Thus, the magnetic polarity of the reeds could be reversed by the application of external electric currents, and the reeds would retain the polarity. In one polarity, the contacts are open, and in the opposite polarity, the contacts are closed. This type of switch is called a remreed switch. It is smaller and more reliable than the ferreed switch that it replaced.

Stored program control

The use of a programmable, digital computer for control in the No. 1 ESS system is called stored program control. Dual processors continually check each other for faults. The programs to control the processors and translation information are stored in a semipermanent memory, called the program store. The status of lines and trunks, called numbers, and other temporary information is stored in the call store. As shown in Figure 7.7, the central control periodically scans the status of lines and trunks over scanners. Signals to operate the switches are sent over distributors.

The No. 1 ESS system sold well, and roughly 1,000 machines were in service. All of its features are controlled by software, and updating that software has become a big chore, mostly because of the sheer size of the programs. Many of the new, functional services (such as call waiting and distinctive ringing) available today from local telephone companies were initially obtained over No. 1 ESS machines.

Digital switching systems

Lucent Technologies No. 4 ESS™ switching system

The No. 4 ESS digital switching system, shown in Figure 7.8, was first installed in the Bell System in 1976. The No. 4 ESS system was the first all digital switching system, utilizing stored program control in a central processor and an all-digital switching network and serving a maximum of 53,760 two-way voice circuits and 550,000 call attempts per peak hour. The No. 4 ESS system is used in toll and tandem switching. The technology used in the No. 4 ESS system is all solid state, using integrated circuits where possible.

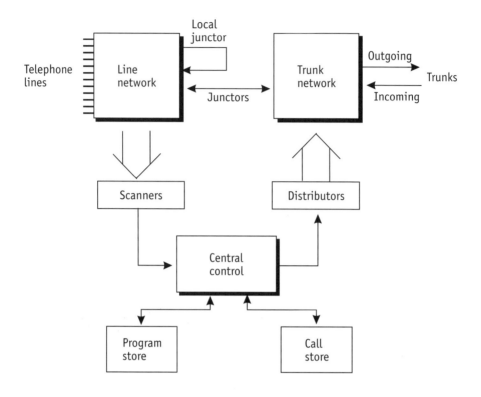

Figure 7.7   Dual processors form central control in the No. 1 ESS system. Access to and from the switches in the switching network is over scanners and distributors.

The No. 4 ESS digital switching system was initially manufactured by AT&T's Western Electric Company. It is now manufactured by Lucent Technologies. Since its initial introduction, the No. 4 ESS switching system has undergone many hardware updates and improvements. The current (1998) No. 4 ESS system can serve a maximum of 64,512 two-way voice or data circuits and over one million call attempts per peak hour.

Time switching in the No. 4 ESS system is accomplished in time-slot interchange units (TSIUs) composed of many buffer memories (see Figure 7.9). Each buffer memory can store 128 time slots. The 128 slots are used for 120 digital voice circuits, and the remaining eight slots are

Figure 7.8   Photograph of the AT&T No. 4 ESS switching system, now manufactured by Lucent Technologies. (Photo courtesy of AT&T Network Systems.)

used for maintenance purposes. Eight elastic stores, including one elastic store as a maintenance unit, comprise the input to a TSIU, as shown in Figure 7.10. Elastic stores are buffer memories that are able to accommodate the different delay requirements for different time slots. The eight elastic stores are assigned uniformly to eight input buffer memories for load-spreading purposes. A device called a decorrelator performs the assignment of the specific slots in specific buffer memories. Each output line, called an A-link, from the buffer memories carries $\frac{7}{8}$th of the input circuits, or $(\frac{7}{8})120 = 105$ circuits. This expansion factor of 105 circuits to 128 time-slots makes the space-switching network virtually nonblocking. Each TSIU also contains a similar, mirror image, output side circuit

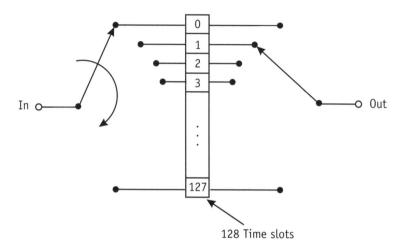

128 Time slots

Figure 7.9   The buffer memories used in the No. 4 ESS system contain 128 time slots with each slot containing 8 bits. The time slots are written into the buffer in sequence, but are read out in a different order determined by the desired connections to be made.

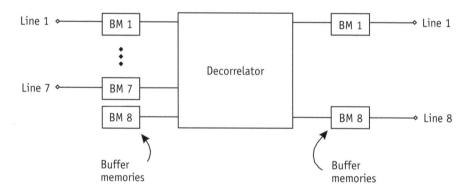

Figure 7.10   A time-slot interchange unit (TSIU) in the No. 4 ESS system consists of eight input elastic stores and eight input buffer memories. A decorrelator spreads the load uniformly across all the buffer memories. The TSIU serves seven digital lines, each carrying 120 digital circuits.

composed of eight more buffer memories and a device called a recorrelator. Each A-link has 128 time slots per 1/8,000th sec from the input buffer memories, and also to the output buffer memories. A single TSIU can serve a total of 7 × 120 = 840 digital circuits. The input and output lines to the TSIU are coaxials.

Hardware upgrades have been introduced since this initial implementation. The current implementation utilizes a single large buffer memory for each TSIU. This allows higher circuit utilization while still maintaining a virtually nonblocking space-switching network. The eight input buffer memories and the eight output buffer memories are all compressed into a single large buffer memory and the first and fourth space-division switches are not needed. The decorrelator and recorrelator devices are also eliminated. The single large buffer memory now contains 4,096 time slots, with 2,048 time slots being reserved for additional features. Of the remaining 2,048 time slots, 1,024 time slots are allocated for input circuits and 1,024 time slots are allocated for output circuits. For each of these two groups of 1,024 time slots, 1,008 time slots are used to carry voice or data circuits, expanding the capacity of each TSIU to 1,008 two-way digital circuits. The remaining 16 time slots of each group are reserved for maintenance purposes.

A total of 128 TSIUs can be connected to a center-stage, time-multiplexed switch (TMS), where the stage 2 and stage 3 space-division switching occurs. The output port of the TSIU input buffer memories connect to the input of the TMS, and the output of the TMS connects to the input port of the TSIU output buffer memories, as shown in Figure 7.11. Because each TSIU has eight output lines, the TMS must have 128 × 8 = 1,024 inputs and another 1,024 outputs for the TSIU input lines. The time-multiplexed switch is a huge, time-multiplexed space switch in the form of a 1,024 × 1,024 matrix switch. Each digital circuit switched through the switching network is one-way, and hence two such one-way circuits are needed to create a two-way circuit. Therefore, the switching network in the No. 4 ESS system can handle a maximum of 1,008 one-way circuits per TSIU times 128 TSIUs, which equals 129,024 one-way digital circuits, or 64,512 two-way digital circuits.

The time-multiplexed switch (TMS) performs its space switching in stages to distribute the load uniformly, as shown in Figure 7.12. In the original implementation, each of the eight 128-time-slot input buffer memories within a TSIU connected to an 8 × 8 time-multiplexed space switch that formed the first stage of space-division switching. This first-stage space switch allowed any of the eight input buffer memories to

Figure 7.11 The overall architecture of the switching network used in the No. 4 ESS system is a basic time-space-time digital switching system. The space switch is a time-multiplexed solid-state coordinate switch with as many as 1,024 inputs and 1,024 outputs. Each cluster of eight inputs or eight outputs is served by a TSIU.

connect to any of the second-stage space switches within the TMS for uniform load distribution. The second-stage space switches within the TMS are capable of handling 256 inputs and 256 outputs. This stage-2 switch is actually composed of sixteen 16 × 16 space-division switches. The 256 outputs of the stage-2 switches connect to 256 inputs on the stage-3 switches via B-links, as shown in Figure 7.12. The stage-3 switches also provide 256 inputs and 256 outputs utilizing sixteen 16 × 16 space-division switches. There are actually four 256 × 256 TMS grids that combine to form a 1,024 × 1,024 space-division switch. The outputs of these four 256 × 256 stage-3 switches connect to the output buffer memories via a fourth-stage space-division switch that is part of each TSIU. Together, these four stages of time-multiplexed space-division switching form a complete 1,024 × 1,024 space switch capable of carrying 128 time-slots per port. This yields a total switch center stage capacity of 1,024 ports x 128 time-slots × 8000 samples/sec × 8 bits/sample = 8.388 gigabits/sec.

In the latest implementation (circa 1996), the input and output buffer memories within the TSIUs are all contained in a single large buffer memory. Dedicated time slots are assigned to support fully all 128 time slots to and from the TSIU stage-2 and stage-3 switches, and thus the stage-1 and stage-4 space switches are not needed. With this arrangement, the blocking associated with these stage-1 and stage-4 switches is also eliminated,

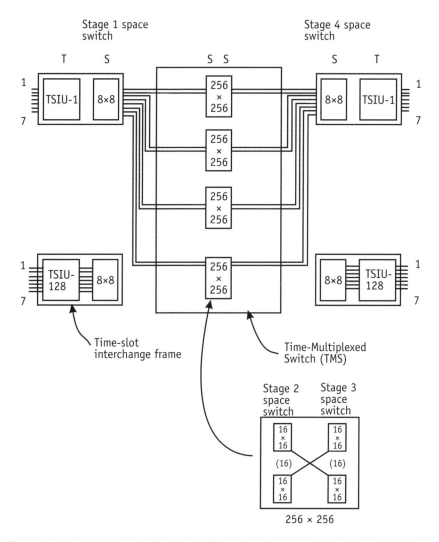

Figure 7.12 Switching in the No. 4 ESS system is accomplished in a TSSSST order. A time-slot interchange unit and an 8 × 8 time-multiplexed space switch form a time-slot interchange unit (TSIU). The time-multiplexed switch (TMS) performs space switching of the 128 TSIUs at the input and output. The TMS is formed from four 256 × 256 switches. Each 256 × 256 switch is comprised of two stages of 16 × 16 switches.

increasing the maximum number of possible circuits from the original 107,520 one-way circuits to the current 129,024 one-way circuits.

Conventional analog circuits need an interface to convert them to the digital format required by the No. 4 ESS system. Many types of equipment perform this conversion. The voiceband interface unit (VIU) digitizes and multiplexes 120 voiceband circuits at 64,000 bps each. D-type channel banks, such as D4 or D5 channel banks, can digitize and multiplex 24 voiceband circuits into standard DS-1 transmission formatted signals, which the No. 4 ESS switch digital interface frame further multiplexes into 120 voiceband circuits for connection with the TSIUs. Other switching systems, such as the Lucent 5 ESS™ switch, described later in this chapter, digitize voiceband circuits directly and pass customer information over DS-1 channels to the digital interface frame. In the current implementation, the expanded time-slot interchange directly terminates DS-3 formatted transmission signals, which carry 672 voiceband circuits over a single facility.

In the No. 4 ESS switching system, 6 or 24 paths through the switching network can be synchronized to set up either 384-kbps or 1.536-Mbps digital connections for wider bandwidth services.

Nortel DMS™ switching system

Nortel Networks (formerly Northern Telecom) manufactures the DMS (for Digital Multiplexed System) series of digital switching systems. A photograph of a DMS machine is shown in Figure 7.13. The DMS-100 family of systems is intended for local offices, and the DMS-200 family is intended for toll and tandem applications. The various machines are all similar in basic structure; the differences are in software and in peripheral modules. The DMS-200 first saw service in 1978 and could handle 61,444 trunks.

The overall architecture of the DMS system, shown in Figure 7.14, is fairly simple and consists of two independent, fully duplicated digital switching networks organized into two planes. In case of trouble, each switching network can take over for the other and can carry the full load. The switching networks interface with lines and trunks through various traffic interface modules. Control is by two independent central processing units (CPUs) with one CPU active and the other standby. The two CPUs operate in synchrony and continuously check each other, although one is in control. Each CPU has its own program store and data store. The

Figure 7.13   Photograph of Northern Telecom's DMS switching system. (Photo courtesy of Northern Telecom.)

actual programs used to control the operation of the system are contained in program store, which contains as much as eight million 17-bit words of memory. Translation information, temporary call information, and other administrative information are stored in the data store. The CPUs communicate with the switching networks through two message switches (MS). Magnetic tape units, visual display units, and teletypewriters are connected to the MS through two input/output message switches (IOM). The functional elements of the control section are interconnected by a flexible bus structure, operating at 128 Mbps, that facilitates multiple processors, memories, and switching architectures.

Virtually every unit in the DMS switching system is duplicated. This is both for load sharing purposes and also so that one unit can take over for the other to achieve reliability in case of failure of one unit. Extremely reliable operation is a major consideration in switching system design, particularly for large toll-switching systems that handle considerable

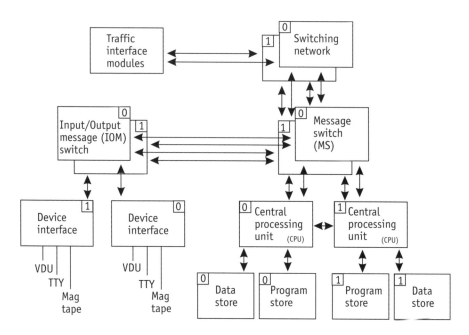

Figure 7.14 The overall architecture of the Northern Telecom DMS digital switching system consists of a switching network organized into two planes, various traffic interface modules, and a fully duplicated common control.

amounts of network traffic. Loss of any traffic immediately translates into a loss of significant revenues.

The switching network is organized into two separate but fully duplicated planes, numbered 0 and 1. The specific architecture for the switching network has evolved over time. In older DMS machines, the switching was performed in two stages of TST switching and later was changed to an STTS architecture. Current DMS machines use optical fiber for input and output from the switching network, called "E-NET" for Enhanced NETwork.

Each fiber in the E-NET carries 512 time slots in 1/8,000 sec, or 125 μsec. The switching network is a software-controlled, single-stage, non-blocking, time switch capable of switching 131,072 one-way digital circuits or 65,536 two-way digital circuits. The switching network, shown in Figure 7.15, consists of eight horizontal buses for input and eight vertical buses for output. Each bus carries 16,384 time slots in 1/8,000th sec.

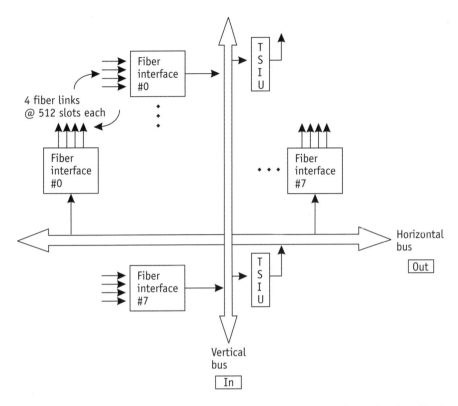

Figure 7.15 The architecture of the E-NET switching network can be visualized as a large, single-stage, nonblocking time switch capable of switching 131,072 one-way digital circuits. The actual switching is accomplished by 64 time-slot interchange units located between vertical and horizontal buses. The DS512 fiber interface connects four fiber links to a bus.

Various interfaces are used to combine signals for input to and output from the buses. The DS512 fiber interface handles four fiber links with each link carrying 512 time slots. Eight such fiber interfaces can be served by one bus. A time slot interchange unit (TSIU) is located at each of the 64 crosspoints between the vertical and horizontal buses. The input to the TSIU comes from a vertical bus, and the output goes to a horizontal bus. Each TSIU stores 16,384 time slots in a double-buffered configuration so that the delay through the TSIU is always a fixed 125 msec for all for all slots. Nortel calls the TSIU a "double-buffered 16Kx16K time switch."

Here, the "K" refers to the binary "K" of 1,024 instead of the decimal "k" of 1,000.

E-NET is a most impressive digital switching network. The processor that controls it has much to do to assure that each time slot makes its way to its destination at precisely the correct time and through the correct time-slot interchange unit. E-NET can be visualized as a single, very large time switch with multiple input links and multiple output links—an approach discussed in a previous section of this chapter.

Analog and digital lines and trunks interface to the switching system through traffic interface modules. The two-way digital output lines, called links, from each traffic interface module connect to both planes of the switching network. The outgoing link connects directly to both planes, while the incoming link has a time-multiplexed switch to choose the appropriate plane. In effect, each two-way link from a traffic interface module sends to both planes and receives from one plane, as shown in Figure 7.16. There are three types of traffic interface modules: a line-concentrating module (LCM) working through a line-group controller (LGC), a trunk module (TM), and a digital trunk controller (DTC). Different modules are configured, depending upon the specific application of the DMS switching system.

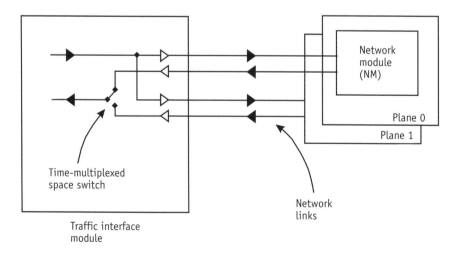

Figure 7.16 A time-multiplexed space switch in each traffic interface module switches between the two planes of the switching network.

We first describe the line-concentrating module. A line-con- centrating module serves 640 analog local-loop circuits or lines. Each local loop is connected to a line card, shown in Figure 7.17, that supplies battery and functions (collectively known as BORSCHT). The line card is where analog-to-digital conversion takes place for each local-loop termination. The output from the line card is a two-way PCM, or digital, signal at 64 kbps. A total of 32 such PCM signals are time multiplexed together by the bus interface card, as shown in Figure 7.18. The bus interface card has 30 output time slots. The two-way digital circuit from the bus interface card is called a terminal link and operates at 2.56 Mbps.

Each of the 32 time slots carried on the terminal link contains 10 bits. PCM voice and data is carried in 30 of the time slots, which means that since there are 32 input circuits, a small amount of concentration takes place in the bus interface card. Each 10-bit time slot contains eight bits of PCM voice and data information along with two bits for signaling and intraswitch communication purposes. For example, one of the two bits indicates whether the line is on-or off-hook for a signal coming from the line card over a line circuit . The remaining two of the 32 time slots carry control information.

Figure 7.17   Each local-loop line is served by its own line card. The line card supplies battery, supervises the loop, gives access to various line tests, performs two-wire to four-wire conversion, and performs analog/digital conversion. The output from the line card is a pulse code modulation (PCM) signal.

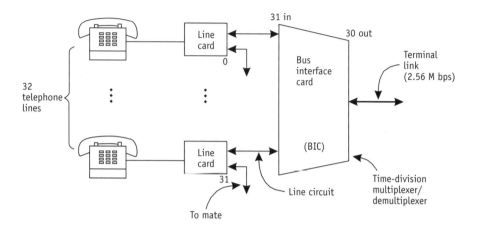

Figure 7.18 The bus interface card performs time-division multiplexing and demultiplexing for the telephone lines served by 32 line cards. The output from the bus interface card is called a terminal link. A terminal link carries a digital signal at 2.56 Mbps.

Twenty bus interface cards, numbered 0 to 19, connect to a time-multiplexed space switch, as shown in Figure 7.19. The multiplexed switch ultimately serves 640 local-loop circuits connected to their line cards and performs concentration of the loop circuits. The input to the time-multiplexed space switch is 20 terminal links, with each terminal link carrying 30 two-way digital voice/data circuits. The output from the time-multiplexed switch is from 2 to 6 channel links, depending on the concentration ratio, with each channel link likewise carrying 30 digital voice/data circuits. For a concentration ratio of 5:1, the 600 input circuits are concentrated onto 120 output circuits carried on four channel links. Each channel link operates at 2.56 Mbps in each direction. As many as 20 channel links connect to a line group controller (LGC), which performs call-processing functions and switching between the two network planes. From 2 to 16 network links, each operating at 2.56 Mbps over copper wire, connect the line-group controller to the switching network. In DMS systems using E-NET, the information in up to twenty 2.56-Mbps channel links is combined together by the line-group

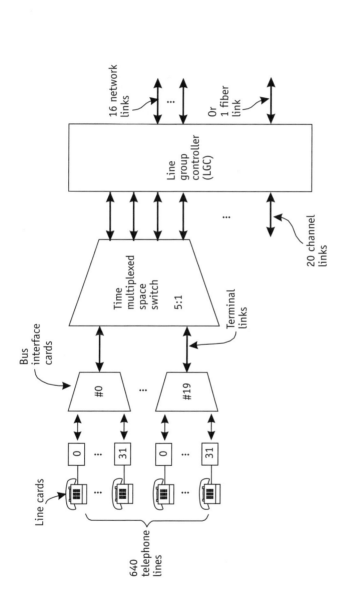

Figure 7.19 A total of 20 bus interface cards are served by a time-multiplexed space switch. For the 5:1 concentration shown here, the 600 input voice-channel time slots are reduced to 120 voice-channel time slots. These 120 time slots are carried over four two-way channel links each operating at 2.56 Mbps. An additional two time slots for signaling and control purposes has been added to each channel link so that each link actually carries 32 10-bit slots. In newer DMS systems using E-NET, a single two-way fiber link (two fibers) connects the LGC to the switching network; in older DMS systems, as many as 16 network links, each operating at 2.56 Mbps over copper wires, connect the LGC to the switching network.

controller and transmitted over a single optical-fiber link carrying 512 time slots (a so-called DS512 signal).

Two line modules operate together as mates. The processor in each line module can access all the cards in the other line module. This is so that in case of trouble with one processor, the other processor can take over, even though the traffic-handling capacity would be reduced. A master processor is responsible for controlling the operation of the line module, and a signaling processor is responsible for controlling line scanning, ringing, and other supervisory and signaling functions.

The DMS system can be used for a wide variety of switching applications, including local, toll, tandem, cellular radio, private branch exchanges, and interexchange. It interfaces with various loop-carrier systems, such as Nortel's DMS-1 urban loop carrier and competing products, and also with digital trunks and with ISDN and ADSL lines.

Lucent 5ESS® switching system

The Lucent Technologies 5ESS digital switching system, shown in Figure 7.20, is intended for a wide variety of applications, such as local, toll, wireless, and PBX switching. The first system, originally manufactured by AT&T's Western Electric, was cut-over in March 1982. As is usual with digital switching machines, the technology continues to progress, and newer machines are made with even greater capacities. For example, the 1983 version of the 5ESS system could handle 50,000 lines and 130,000 telephone calls per hour. In 1998, the 5ESS-2000 switch, fully configured, could handle over 200,000 lines and 45,000 trunks simultaneously and support 1,000,000 telephone calls per hour. The 5ESS switching system will undoubtedly continue to evolve in the future as new features and interfaces are added.

The overall architecture of the 5ESS system, shown in Figure 7.21, consists of an administrative module (AM), a single, central communications module (CM), and various switching modules (SMs). The switching modules are time-slot interchange units (TSIUs) that perform time switching. The communications module is a time-multiplexed space switch that performs space switching between the switching modules. The communications module also performs message switching between the switching modules and to the administrative module. The AM is

Figure 7.20   Photograph of the 5 ESS switching system, currently being manufactured by Lucent Technologies. (Photo courtesy of AT&T Network Systems.)

responsible for such functions as billing, call translation, trunk routing, and other administrative features. Control is distributed throughout the switching system, and each switching module has its own processor and memory for module control. The architecture is distributed and is modular so that custom configurations can be designed for specific installations.

The switching module communicates with the time-multiplexed switch (TMS) in the communications module over network control and timing links. These links are two pairs of multimode fiber. Each fiber carries either 256 or 512 time slots in 1/8,000 sec, with one slot of a frame allocated for control messages. Each time slot consists of 16 bits with 8 bits for digital voice and data and the remaining 8 bits for signaling and control, such as on-off hook indication, the status of the slot, and parity. The links used by the SM-2000 switching module carry 512 time slots and operate at 65.536 Mbps. Older switching modules carry 256 time slots and operate at 32.768 Mbps. Switching modules can be located remotely

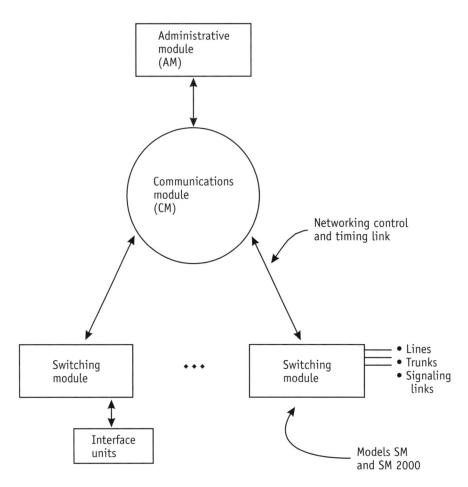

Figure 7.21   The overall architecture of the Lucent 5ESS digital switching system consists of many switching modules that interface to analog and digital lines and trunks and perform time-slot interchange.

and communicate with the communications module via digital trunks operating over copper wire or optical fiber at distances up to 600 miles. Switching modules can operate alone or in clusters if they lose contact with the central office, using dedicated trunks for direct access to the network.

The TMS is organized into two planes: one plane serves 256 odd-numbered time slots, the other serves 256 even-numbered time slots.

The TMS is a high-speed matrix switch with a maximum of 192 ports capable of serving 190 SMs. The information in a time slot passing through a time-slot interchange unit can be delayed a maximum of 125 ms. The TMS, however, has zero delay. Each plane of the TMS must reconfigure its switching matrix 256 times every 1/8,000th sec. That works out to 2,048,000 reconfigurations per second. The communications module contains the network clock that synchronizes the total operation of the switching system.

Shown in Figure 7.22, the switching module is primarily a TSIU with various interface units connecting to analog and digital lines and trunks and signaling links. As shown in Figure 7.23, a module control unit supervises the operation of the switching module; each switching module has its own control unit, and thus control is distributed throughout the system. The optical signals carried over the fiber links are converted to electrical signals by dual-link interfaces (DLIs) for an SM and quad-link interfaces for an SM-2000 in the switching module and the communications module. The various interface units in the switching module are connected to the TSIU over peripheral interface data buses (PIDBs). Each PIDB consists of several serial data circuits, with each individual circuit carrying 32 16-bit time slots. Two time slots are used for control information by the processor that controls the switching module. The switching modules can be equipped with a choice of several microprocessors and a range of memory configurations.

A variety of interface units are available for the switching modules. The subscriber line unit (LU), shown in Figure 7.24, interfaces with local-loop telephone lines. A single LU serves from 256 to 640 lines, depending upon the desired concentration for specific traffic. The lines are concentrated and accessed by a space-division matrix, or crosspoint, switch. The concentrator has 64 output lines, and thus 640 telephone lines can be served with a 10:1 concentration, 512 lines with 8:1 concentration, 384 with 6:1 concentration, and 256 lines with 4:1 concentration. Each two-wire output line from the concentrator is connected to a channel circuit that supplies battery, performs two-wire-to-four-wire conversion, supervises the status of the loop, and performs analog/digital conversion. The integrated services line unit (ISLU) supports the basic rate interface (BRI) used for ISDN lines. The ISLU uses digital

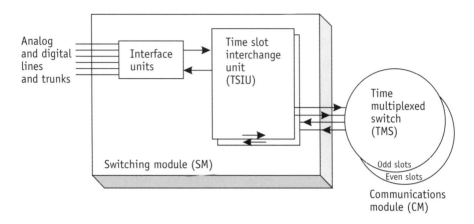

Figure 7.22   The time-multiplexed space switch (TMS) in the communications module is organized into two planes: one for odd-numbered time slots and the other for even-numbered time slots. The switching module consists of a bidirectional TSIU. Various interface units connect to analog and digital lines and trunks.

Figure 7.23   Each switching module has its own module control unit. The signals sent over the optical links between the switching module and the communications module are converted to electrical signals by dual-link interfaces located at each end of the fiber links. Peripheral interface data buses (PIDB) carry signals between the interface units and the TSIU.

concentration, as opposed to the analog concentration used in the subscriber line unit.

The digital line trunk unit (DLTU), shown in Figure 7.25, interfaces with DS1 digital trunks at 1.544 Mbps, usually T1 facilities. The digital facility interface (DFI) handles a single DS1 trunk and converts the bipolar pulses of the DS1 to the unipolar pulses used in the 5ESS system. The 24 time slots in the DS1 are distributed over 32 time slots that are sent over the peripheral interface data bus: 24 slots carry voice/data information and 8 slots are evenly distributed as empty fill. The DLTU can have as many as 10 DFIs. In Europe, a digital trunk operates at a rate of 2.048 Mbps and contains 32 time slots, and in this case, the DLTU can have only as many as 8 such DFIs. The trunk unit (TU) interfaces with 64 analog trunks and converts their analog signals into 64 time slots.

The access interface unit (AIU) is a new platform designed to serve a variety of access lines, including basic analog telephone lines and ISDN lines. The AIU supports up to 640 lines per shelf and can be expanded in 32-line increments. A fully configured AIU cabinet will support 3,584 lines, twice the capacity of the analog line unit. The AIU is nonblocking and does not concentrate traffic, thereby being particularly suited for Internet access.

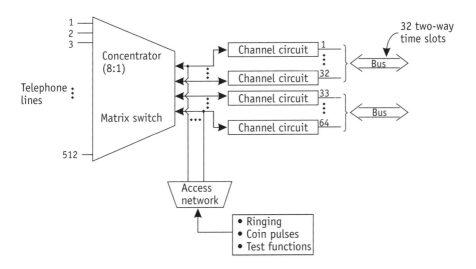

Figure 7.24   The subscriber line unit can serve from 256 to 640 telephone lines, depending upon the acceptable amount of concentration. An individual line is accessed by the concentrator, which is a space-division matrix switch using solid-state gated diodes at the crosspoints.

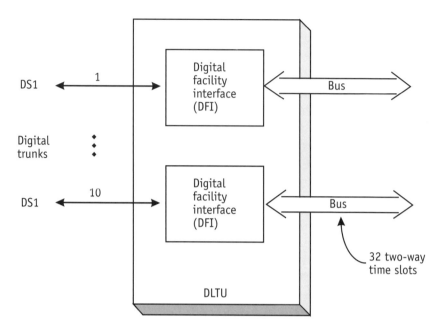

Figure 7.25 The digital line trunk unit (DLTU) connects DS-1 digital to the data buses through digital facility interfaces (DFIs).

The packet switching unit (PSU) supports packet data switching, packet signaling messages, and wireless CDMA and TDMA vocoders. The digital network unitSONET (DNU-S) is a peripheral interface unit to terminate SONET trunks. The integrated digital carrier unit (IDCU) interfaces to digital subscriber loop carrier (SLC) systems and terminates up to 40 DS-1 facilities. The IDCU supports conventional and ISDN lines.

Private branch exchanges

Many organizations have a need for their own exclusive switching systems, which are in effect a telephone switching exchange acting as a branch of a central office for the private use of the organization. Such switching systems are called private branch exchanges, or PBXs for short. The PBX usually is placed on the premises of the organization.

The various technologies used for PBXs closely parallel the development of the switching technologies and systems used for telephone-network switching. The earliest PBXs were manual switchboards operated by human attendants, usually situated near the entrance to the facility so the operator could also serve as the receptionist. The attendant knew the comings and goings of all the staff and could take messages and forward calls as needed. Cordless switchboards were developed to ease the work of the attendant.

The role of the attendant was automated through the use of such electromechanical technologies as step-by-step and crossbar, adapted for use in PBXs. Such automated PBXs are called private automatic branch exchanges (PABXs). Time-division switching utilizing digital technology and computer control has created a whole new generation of PBXs, called computerized branch exchanges (CBXs). The term PBX is still used to refer generically to any type of private branch exchange, be it automatic, electronic, or computerized.

Some of the newest PBXs utilize time-division switching of pulse-code modulated (digital) speech signals. Special telephone instruments that perform analog-to-digital conversion are used and transmit digital signals over the local on-premises wiring. The digital signal can also represent text, and so these types of all-digital PBXs are particularly applicable to electronic mail and remote access to computerized data bases. Since the speech signals have already been digitized, such PBXs frequently use T1 trunks for access to the local central office and the public switched network.

Modern PBXs can be quite large, serving thousands of telephone extensions at a single location. Many large companies have locations scattered about the country, and therefore it is quite natural to connect together all the PBXs serving the locations through a private long-distance network. The result is a corporate communications network that can be quite complex and extensive. Synthesizing together many different PBX vendors and network providers to create an integrated telecommunication facility is the job of today's corporate telecommunication manager.

Most PBXs allow individual telephones to directly dial outside over the public switched network, a feature that is called direct outward dialing (DOD). However, if the local telephone company assigns only a single telephone number to the PBX, then all inward-coming calls must be answered by an attendant who makes the final connection to the desired extension. An alternate approach is the use of recorded speech asking the calling party to touch-tone enter the extension number so that the PBX itself can automatically make the final connection. An even better solution is for the local telephone company to assign an entire three-digit exchange number to the PBX so that all incoming calls can directly reach any known extension simply by dialing the complete 7-digit number. This feature is known as direct inward dialing (DID).

PBXs have their advantages. Equipment is not controlled by the local telephone company. The PBX user is in complete control of their telecommunication system and can do as they wish, when they wish without waiting for the telephone company to be available. PBX equipment can be purchased or leased.

For a large PBX installation, the telecommunication manager in reality is operating a small telephone company with all the inherent problems. Another solution is the use of the switching system at the local central office as if it were a private branch exchange. This type of service, provided by the local Bell telephone company, is called Centrex service. It is offered on a leased-service basis and competes with the purchase of an on-premises PBX. Centrex service operates with standard telephones, and instructions are indicated by momentarily operating the hook, called hook flashing, which many users find confusing and disconcerting.

Telemarketing involves the use of the telephone, and usually a toll-free 800 number, for placing orders for goods and services. An agent answers the call and types the order into a computer. Many agents are available to answer the calls and a means is necessary to distribute the calls uniformly to available agents. This function is called automatic call distribution (ACD) and is performed by an automatic call distributor installed at the PBX or at the local office.

If one has only a small number of extensions, a simple key telephone system (KTS) will usually suffice. Keys on each telephone are depressed to seize a line. Typical key telephones have six keys, offering access to five lines—with the sixth key a "hold" key. The keys are lighted and flash in different patterns to indicate whether the particular line is in use, on hold, or ringing. Older key systems utilizing electromechanical relays and multiwire-pair cables are being replaced by newer electronic systems that mimic the features of key telephone systems but work over only a few pairs of wire and utilize computer technology.

BORSCHT

A local-exchange switching system that serves local-loop lines is responsible for performing a number of functions that are lumped together under the acronym BORSCHT.

The local switching office must supply battery feed at 48V dc to power the telephone instruments connected to the local loops. The switching equipment must be protected from high voltages that might occur during a lightning strike on the line, and hence some form of over-voltage protection must be supplied. The local office supplies the ringing signal necessary to cause the telephone instrument to ring. The local switching machine needs to supervise whether the local line is in use or not, and hence supervision must be supplied. In the case of a digital switching machine, the analog signal on the local loop needs to be converted to a digital signal and vice versa, and hence coding and decoding must be supplied. The two-wire local loop needs to be converted to the four-wire signal required for modern switching systems, and hence a hybrid function must be performed. Last, various type of loop testing are needed.

Putting all this together, we have:

Battery feed;

Overvoltage protection;

Ringing;

Supervision;

Coding and decoding;

Hybrid; and

Testing.

Switching—an assessment

The world of telecommunication has gone digital. We saw in Chapter 5 how optical fiber and time-division multiplexing have driven transmission toward digital. In this chapter, we saw how time-division switching and computer control have driven switching toward digital.

Voice communication is an analog affair. Hence, there will always be a need for analog-to-digital conversion. Most telephone instruments and, of course, the local loop are analog. Every local loop travels all the way back to the central office, thereby contributing to all the wire that is found there. Optical fiber can multiplex together hundreds (or more) of telephone conversations and can eliminate the need for a physical wire pair for each conversation. Some central-office switching systems have switching modules that can operate remotely from the central office. Coupling these remote modules with fiber creates a new local-network topology in which only the last few hundred feet to each telephone is wire pair. The wire pairs are brought to a remote switching module for switching and multiplexing, as shown in Figure 7.26. If this trend were to continue, the role of the central office as a local switching point will surely change, and the central office will offer mostly access to interexchange carriage.

The newer digital switching systems use optical fiber to carry signals between interface modules and the switching network. The conversion of light signals to electrical signals would be an unnecessary chore if light signals could be directly switched on fiber. So-called photonic switching is an area of continuing research and might indeed find application in switching machines of the distant future.

Many telephone companies invested heavily in replacing step-by-step and crossbar systems with AT&T's No. 1 ESS system. The No. 1 ESS system is a computer-controlled space-division switching system. The newer systems, such as the Lucent 5ESS system and the Nortel DMS system, are computer-controlled time-division switching systems. Since most trunks are digital, these newer digital switching systems interface easily with digital trunks. They also are physically smaller than the No. 1 ESS machine. Hence, No. 1 ESS machines have been replaced with the newer all-digital machines, even though the older No. 1 machines were

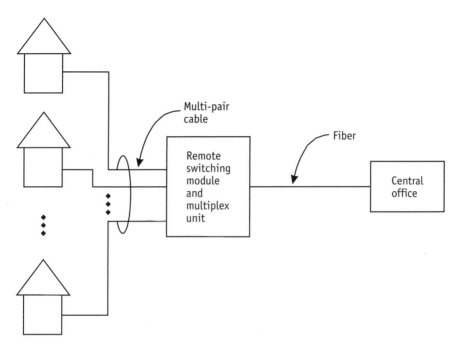

Figure 7.26 Remote switching modules and multiplex units are changing the topology of the local network.

still functioning perfectly and had years remaining on their useful service life.

Electromechanical switching automated work previously performed by human operators and allowed telephone service to be universally affordable by everyone. However, the electromechanical technology was inflexible and did little more than switch telephone conversations. The functionality provided by human operators was lost. No longer would calls be forwarded, or messages be taken, or calls be interrupted. Now, with computer control and the "intelligence" that it provides, flexibility and functionality have returned again to telephone service. Such services as call forwarding, call interrupt, voice mail, and calling-number identification are again available, but at a price easily affordable to everyone.

Selected bibliography

Hayward, W. S., Jr. (ed.), "The 5ESS Switching System," *AT&T Technical Journal* (special issue), Vol. 64, No. 6, Part 2, July/Aug. 1985.

Hills, M. T., *Telecommunications Switching Principles*, Cambridge, MA: The MIT Press, 1979.

Holland, H. G., et al., "The 5ESS-2000 Switch: Exceeding Customer Expectations," *AT&T Technical Journal*, Nov./Dec. 1994, pp. 28-38.

Terry, J. B., D. R. Young, and R. T. Matsunaga, "A Subscriber Line Interface for the DMS-100 Digital Switch," *IEEE National Telecommunications Conference (NTC) Digest*, 1979, Vol. 1, pp. 28.3.1–28.3.6.

Terry, J., H. Krausbar, and J. Hood, "DMS-200 Traffic Peripherals," *IEEE Communications Conference Record (ICC'78)*, 1978, pp. 32.3.1–32.3.5.

"No.1 Electronic Switching System," *Bell System Technical Journal*, Vol. XLIII, No. 5, Parts 1 & 2, Sept. 1964.

"No.4 ESS," *Bell System Technical Journal*, Vol. 56, No. 7, Sept. 1977.

8

Signaling

Introduction

Telephone service involves a vast network of transmission and switching facilities that are controlled by the telephone user. In order to accomplish this, signals of various kinds are sent over that network to control its operation and to indicate its status. The topic of signaling deals with those control and status signals. Some of the information sent over signaling facilities is also used for billing purposes.

Manual exchange signaling

One way to understand some of the functions of signaling is to consider manual switching by human operators.

In the early days of telephony, exchange service was accomplished with manual switching by a human operator. The calling subscriber first alerted the operator by turning the crank on the telephone, which caused a lamp to flash on the panel at the exchange office. The operator saw the flashing lamp and plugged in on that line. The calling subscriber then verbally gave the operator the name or number of the called party. The operator then visually checked the lamp associated with the called party's line to determine whether the line was available. If it were not, the operator verbally informed the calling subscriber that the called party's line was in use. If the line were available, the operator rang the called party's phone and made the final connection of the two parties. Lamps indicated when a party hung up, and the operator would then take down the connection.

If the call were from one exchange to another, the operator at the calling party's exchange used special interexchange lines, called trunks, to reach an operator at the called party's exchange. The number being called was passed verbally from operator to operator, and the operator at the called party's exchange made the final connection.

Some of the signaling functions that occurred during this manual connection were the alerting of the operator that service was desired, the verbal specification of the called party's identity or address, the supervision by lamps of the status of the telephone lines or circuits, and the reporting by the operator of the progress of the call to the calling party.

Clearly, the making of a telephone connection during the early days of telephone service involved a large amount of human labor. Over the years, technology reduced and finally eliminated all human labor in making a telephone connection. This was accomplished through automated switching machines and various electrical signals to request service, forward telephone numbers, and make the actual connection of circuits.

Functions

The four general functions of signaling in modern telephony are as follows: (1) alerting, (2) transmitting address information, (3) supervision, and (4) transmitting information.

Alerting deals mostly with requests for service, facilities, and action of some kind. The initial request by the subscriber for service is an alerting function. The local office might send alerting signals to other offices requesting interoffice trunks. A local switching system alerts the called party to answer the telephone.

The telephone number, or address, of the called party must be transmitted from the calling party to the local switching machine. This is accomplished either by dial pulses or by tones. The switching machines in the network need to have the address so that the call can be completed. These are some examples of the transmission of address information.

Switching machines need to know whether lines and circuits are idle or are in use. They also need to know when a seized circuit is no longer needed and can be released for reuse. The status of circuits and lines thus needs to be supervised. Information signals, such as dial tone, busy tone, and various recorded announcements, must be transmitted to the calling party.

Subscriber-loop signaling

Two major realms of signaling are subscriber-loop signaling and interoffice signaling. We first describe how the four functions of signaling are accomplished using the subscriber's local loop.

The alerting function of requesting service occurs when the handset is lifted from its cradle, or goes off-hook, thereby causing a direct current to flow through the loop. This flow of direct current is sensed by the switching equipment at the central office and is interpreted as a request for service. The flow of direct current continues as long as the telephone is in use, and by supervising this flow, the switching equipment senses that the line is still in use. When the flow of direct current ceases, the switching equipment knows that the conversation has ended. Thus, direct-current (dc) signaling on the subscriber loop is used for alerting and supervisory functions.

Address information can be sent over the subscriber loop in two ways. The flow of direct current can be interrupted by a switch in the telephone dial to generate dial pulses that are counted by equipment in the switching machine at the local office. Dial pulses occur at a rate of about 10 pulses per second. The second way for transmitting address information over the subscriber loop is in the form of unique combinations of two pure tones, called touch-tone dialing.

Information is transmitted over the subscriber loop either as audible tones or as recorded announcements. Four major audible tones are dial tone, line-busy tone, ring-back tone, and trunk-busy tone. Four different frequencies (350, 440, 480, and 620 Hz) are used, either singly or in combination with each other. Distinctive timing patterns are also used.

Dial tone is a continuous signal formed by the simultaneous transmission of a 350-Hz tone and a 440-Hz tone. Such pure tones are also known as sine waves. Ring-back or audible ringing is the ringing signal heard by the calling party. The ring-back signal is formed by the simultaneous transmission of a 440-Hz tone and a 480-Hz tone. The combination is on for two seconds and off for four seconds and then repeats. The line-busy signal is formed by the simultaneous transmission of a 480-Hz tone and a 620-Hz tone. The combination is on for 0.5 second and off for 0.5 second and then repeats. The trunk-busy signal indicates that either switching equipment or transmission facilities are not available to complete the call, and the call should be retried. The trunk-busy signal is also known as the reorder signal. It is composed of the same tones as the line-busy signal but is repeated at twice the rate, namely, 0.25 second on and 0.25 second off.

Another alerting signal that is sent over the subscriber loop is the ringing signal that causes the called telephone to ring. The ringing signal is a tone of 75V (rms) at a frequency of 20 Hz. The ringing signal is on for two seconds and off for four seconds, and then repeats this pattern until the called telephone answers the call.

Interoffice signaling

The oldest and most basic form of signaling to indicate the availability of interoffice trunks was dc signaling. The presence or absence of dc on a trunk indicated whether the trunk was idle or in use. Normal and reverse polarity of the dc was sometimes also used for interoffice-trunk signaling.

Long-distance circuits carry only alternating current (ac), and hence dc signaling could be used on such trunks. Thus, a single-frequency (SF) tone was used, either within the voice band (200 to 3,400 Hz) or outside the voice band (3,700 to 3,825 Hz). One popular scheme was the use of a 2,600-Hz single-frequency tone to indicate the availability of a trunk. The 2,600-Hz tone was placed on idle trunks. Once a trunk was seized, the tone was turned off by the near office, and this signaled the distant office that a call was about to be placed over that trunk. Address information was then sent over the trunk to the distant office, using multiple-frequency key pulsing (described below). When the called party answered, the distant office seized a return trunk and turned off its 2,600-Hz tone. The two-way voice connection was thus established, and billing and conversation began.

The address information was sent over the seized trunk by using combinations of two tones sent at a rate of 10 combinations per second. The tones were at frequencies of 700; 900; 1,100; 1,300; 1,500; and 1,700 Hz. Using all the possible two-frequency combinations of these tones, it was possible to encode all ten digits and six control functions. This type of in-band signaling was called multiple-frequency key pulsing (MFKP).

Though quite simple, in-band signaling had problems. For one, some dishonest subscribers discovered that they could generate their own tones for the fraudulent purposes of avoiding toll charges. For another, some speech signals could cause accidental disconnections. Out-of-band signaling avoided these problems but reduced the usable bandwidth for the speech signal. All these problems are avoided with common channel interoffice signaling, which has replaced totally the older forms of interoffice signaling.

Common channel interoffice signaling

In 1976, a new interoffice signaling scheme was first installed in the Bell System, and, though undergoing steady enhancement, it is currently in use on virtually all interoffice trunks circuits. The new system is called common channel interoffice signaling, or CCIS for short.

With CCIS, as shown in Figure 8.1, a separate circuit between the offices is dedicated solely as a data link for the transmission of signaling information. No signaling information is sent over the voice circuits. The switching machines used in most central offices and in all toll offices are controlled by their own internal digital computers, or processors, that are responsible for the actual switching of the connections. CCIS allows these processors to communicate digitally with each other and to exchange such information as trunk availability and addresses.

Figure 8.1   Common channel interoffice signaling uses separate, dedicated trunks for signaling. Initially, these CCIS trunks operated at 2.4 or 4.8 kbps over analog trunks. Today's common channel signaling (CCS) uses a dedicated 56-kbps digital trunk for signaling. Initially introduced by AT&T for long-distance calls, nearly all local lines today are served by common channel signaling using SS7.

A single analog circuit carries the digital data for signaling purposes. Conventional full-duplex modems operating at either 2.4 kbps or 4.8 kbps were used. Each signaling circuit controls about 1,800 or 3,600 voice circuits, respectively. A newer form of common channel signaling uses digital circuits at 56 kbps, and is described below.

Because the signaling information is transmitted over a circuit separate from the voice circuit, there is a need to determine whether the transmission quality of the voice circuit is acceptable before it is connected for use. This is accomplished by performing a transmission quality check on each voice circuit before it is connected for service. The voice circuit is looped back on itself, and a tone is sent over the circuit. The return level of the tone is checked to be certain it is within specifications.

With conventional signaling over long-distance circuits, a busy tone is generated at the distant office and is sent all the way back to the office near the calling party. This tied up a full voice circuit. With CCIS, a data signal is sent back over the data path, and the busy tone is generated at the near office, hence freeing voice circuits for actual conversations.

The original CCIS system has been enhanced to allow messages (called "datagrams") to be sent to specified destinations in the network switching system. The signaling links were upgraded to 56-kbps digital circuits between the switching offices.

The enhanced CCS system is called the common channel signaling 7 (CCS7) network. It is also sometimes known as signaling system 7 (SS7); outside of North America, it is CCITT 17. The CCS7 system as implemented by AT&T supports integrated service digital network (ISDN) features, advanced 800 service (such as time-of-day routing), and software-defined network (SDN) features. Instructions for network routing along with other information is stored in common databases that are accessed over the CCS network by the switching machines.

Common channel signaling was first installed in AT&T's long-distance network. Common channel signaling is now used between local offices, where it allows for more efficient use of trunks and also supports such new services as calling-number identification. Nearly all local lines are served by SS7 in the United States.

Calling-number identification

Common channel signaling has the capability of transmitting the calling party's telephone number to the distant office. Local telephone companies are using their common channel signaling systems to transmit the calling party's number from office to office. This means that the local office knows the telephone number of the calling party. That number can be sent over the local loop to the premises of the called subscriber. This service is known as calling number identification, or calling number delivery (CND), or caller ID. Telephone companies charge extra to provide the service.

The telephone number of the calling party is sent over the local loop as an in-band signal during the first and second rings of the ringing signal.

It is sent as a short burst of digital data at a rate of 1,200 to 1,800 bps. Phase-shift keying (PSK) is used to encode the digital data for transmission. The digital signal is decoded by circuitry at the customer's premises and is displayed on a unit close to the telephone. The called subscriber either recognizes or does not recognize the number and then decides whether or not to answer the call. Undoubtedly, newer customer premises equipment will include a directory of the names of frequently received numbers and will display the name, if included in the directory, of the calling party. Perhaps with speech synthesis, the telephone of the future will not ring but will announce the name of who's calling! Of course, that assumes that the person calling from the number is the expected person.

The subscriber to calling number identification usually also receives the capability to block certain incoming calls. Simply by entering a code, calls from a specified telephone number are forever blocked until the subscriber enters a code to release them. Clearly, crank and annoyance calls are easily traced with calling number identification. In fact, many telephone subscribers can enter a special code, and the telephone number of the party who just called will be saved and sent to the authorities as an annoying call. Calling number identification is particularly useful to businesses, since with calling number ID they automatically have the telephone number of the calling party and can use that number to pull up the appropriate customer record from a computerized database.

However, many people are concerned that calling number ID is an invasion of their privacy. What does the poor spouse do when calling home late at night "from the office" that is really the local bar? Many people feel that their telephone number is private and should only be released with their specific permission. Thus, voices of discontent have been expressed toward calling number ID, and some states have issued elaborate ways for subscribers to block the forwarding of their telephone numbers. However, should such blocking be the norm or should it be only done on a specific request basis? And will those who subscribe to calling number identification simply block all unidentified calls? Will new customer premises equipment be invented that requires all incoming calls to enter their telephone number or some access code before the called customer's phone will ring?

Calling number identification, call block, and call trace are only some of a number of new services made possible by the custom local area signaling system (CLASS^{SM} service). Some other CLASS services are distinctive ringing in which the phone rings in a different pattern for calls from certain preselected numbers, selective call forwarding in which only certain preselected calls are forwarded to another number, and return call in which the last incoming call can be recalled simply by entering an appropriate code. These types of services are sometimes said to be a feature of the "intelligent" network that has evolved because of the use of computers to control signaling.

Telephone numbering plan

During the early days of telephony, a total of 10,000 lines was the maximum number of lines served by a telephone exchange. Thus, a four-digit number was necessary to specify the party in a specific exchange. The exchange was specified by two alphabetic characters followed by a single digit—for example WA5 for WAverly-five. Three-digit area codes were later introduced to specify the area of the country to be reached. An area code is more technically called a numbering plan area (NPA). Today, we also have country codes for international dialing.

A special nomenclature is used in specifying the overall structure of the telephone numbering plan. The symbol N is used for any of the decimal digits 2 through 9; the symbol X for any of the decimal digits 0 through 9; and 0/1 for the digits 0 and 1 only.

The standard format for telephone numbers in the United States used to be N0/1X-NXX-XXXX. The first three digits, N0/1X, specified the area code or NPA; the next three digits, NXX, specified the local exchange; and the last four digits, XXXX, specified the specific subscriber line in the local exchange. Because the number of area codes possible with the N0/1X format was exhausted, a new format of NXX was introduced for the NPA, starting in 1995.

The format N11 is used for special services. Two examples are: 411 for directory assistance and 911 for emergencies.

The growth in telephone numbers has been so steep that the area codes in many metropolitan areas of the United States have run out of available telephone numbers and new area codes had to be created for different geographic regions. This is a great inconvenience to the subscribers who have to remember their new area code and to businesses who have to print new stationery. In some cases, a telephone call across the street involves dialing ten digits because across the street is in a different area code. An alternative to splitting geographically an old area code into several new ones is to overlay the new area codes and assign them only to new telephone numbers. The problem with such overlaying is that telephone numbers in the same geographic area can have different area codes, which means that everyone must dial ten digits to reach anyone.

Local-loop signaling design

The local loop is still mostly copper wire throughout most of the world. The appropriate gauge of wire must be chosen to maintain the proper flow of direct current in the loop.

The flow of direct current along the local loop operates the line relay at the central office, and in this way, a request for service is initiated. The amount of current must be sufficient to operate the line relay. However, if the resistance of the local loop is too high, not enough current will flow to activate the relay. The resistance of the local loop depends on the total length of the loop and on the gauge of the wire. The resistance of various gauges of wire is as follows:

- 26 gauge—83 ohms per 1,000 ft.;

- 24 gauge—53 ohms per 1,000 ft.;

- 22 gauge—32 ohms per 1,000 ft.; and

- 19 gauge—17 ohms per 1,000 ft.

The maximum allowed resistance for the local loop can be calculated as follows. The telephone instrument requires about 23 mA of direct current for reliable operation of the carbon granule transmitter. The common battery at the central office has a nominal electromotive force of 48V. Thus, the total resistance of the whole circuit (telephone plus local loop plus central office) must not exceed $48/0.023 = 2,100$ ohms. The resistance of the telephone is equivalent to 400 ohms, and the central office equipment likewise has a resistance of 400 ohms. Hence, the resistance of the local loop must not exceed $2,100 - 800 = 1,300$ ohms. This maximum loop resistance of $1,300$ ohms determines the wire gauge for a given loop length.

Area code proliferation—an assessment

As the growth in extension lines, wireless phones, and Internet access continues, the area code problem will only worsen. People should not have to remember numbers to reach another person by phone. The World Wide Web works because the uniform resource location is easy to understand and remember. If I want to reach AT&T by the Internet, I know AT&T's address is www.att.com. I do not remember AT&T's telephone number at its headquarters in Basking Ridge, New Jersey. Nor do I remember my aunt's area code anymore, since it has changed a few times already. We need an entirely new way to reach people by telephone— perhaps something more similar to the method of the Internet. If I want to reach AT&T's headquarters by telephone to speak to a human there, perhaps I enter www.att.com[voice] on my phone and the appropriate number is looked up in a database by the signaling system. For certain, the way we are doing telephone numbers today is archaic and most certainly cannot last much longer. This is yet another example of why an integrative systems approach to telephone service is so essential.

Selected bibliography

John J. L., and P. L. Oxley, "Common Channel Signaling Network Evolution," *AT&T Technical Journal*, Vol. 66, Issue 3, May/June 1987, pp. 13–20.

9

Wireless Telephone Service

Introduction

Wireless cellular mobile telephone service is a high-capacity system for providing direct-dial telephone service to automobiles, and other forms of portable telephones, by using two-way radio transmission.

Cellular mobile telephone service was first made available in the top markets in the United States in 1984, and in a very short time has achieved considerable growth and success. During its first four years in the United States, from 1984 to 1988, it experienced a compound annual growth of more than 100 percent. By 1990, its subscribers in the United States numbered 5.3 million; by 1996, its subscribers numbered over 44 million. Cellular mobile telephone service is also a great success in Europe and Scandinavia, where growth rates rival, and in some cases surpass, those in the United States.

Cellular telephone service was initially targeted at the automobile market, but small portable units have extended the market to nearly everyone on the move with a need to telecommunicate and now includes small personal units that can be carried in a pocket. One wonders whether a wrist radio telephone is only a matter of a few more years. The cellular principle has been suggested on a very low power basis to create community systems that could bypass the copper wires of the local loop—so-called wireless local loop (WLL).

The basic principles of wireless cellular telecommunication are described in this chapter along with a discussion of the various technological aspects of a wireless system that must be specified. These basic concepts serve to introduce the descriptions of the more popular specific implementations, such as the Advanced Mobile Phone Service (AMPS) and the Global System for Mobile (GSM) communication. AMPS is described in some detail in this book, since it was the first system to be implemented and also because its details serve to illuminate the differences between it and other, newer systems. AMPS of the 1980s is based on the analog technology; GSM and many other newer systems of the 1990s are based on digital technology. Standards are a big issue with cellular wireless telephone service, and many different countries have each adopted their own unique approach, with even different systems in use in the same country, as in the United States.

Wireless telecommunication is also known generically as personal communication service (PCS), although the term PCS is sometimes meant to refer exclusively to newer all-digital systems operating at 1,900 MHz. The term PCS has also meant the overall concept of being able to reach a person anywhere at anytime by any means of telecommunication. Because of all these different interpretations, to eliminate any confusion, the term PCS is avoided in this book.

The original AMPS cellular system and a newer digital version operate in a range of radio frequencies from 824 to 894 MHz. Since the demand for wireless service was so great, in the mid-1990s the FCC allocated additional frequencies in the 1,900-MHz band for wireless service and then auctioned off the channels to the highest bidder. The channels were assigned to create six blocks of channels for as many as six separate service providers. The prices paid in the auction by many were far too high to be commercially viable, although in theory, the government made billions of dollars. By the end of the 1990s, a number of the firms who had bid successfully had filed for bankruptcy, with the government collecting little. The lesson learned is that auctions are not the best way to allocate spectrum. Spectrum should probably be leased on a yearly basis with recurring fees.

A form of local or neighborhood two-way radio service, called a personal communication network (PCN), has been proposed. There are also cordless telephones that work over about 100 feet and enable people to walk about the home talking on the telephone. These cordless phones use two-way radio and connect to the standard telephone line. Modern cordless telephones utilize digital technology and, though fairly sophisticated, do not support the handoff feature that makes cellular mobile telephone service such a success.

This chapter describes the more popular wireless cellular mobile communication systems and ends with a description of the various new satellite telecommunication systems that are being developed and introduced. But first, a short history of the development of cellular service in the United States.

A short history

Prior to cellular mobile telephone service, mobile telephony was provided by conventional two-way radio, which allowed only a few dozen two-way radio channels in a given service area. A single, centrally located, high-power radio transmitter served a whole area about 50 miles in diameter.

The service in an area was provided either by a wireline common carrier (such as a Bell company), an independent telephone company, or a radio common carrier. The very small number of users who could be served in a given area meant that the service was quite costly, and the limited capacity of the service meant that many potential customers went unserved. Cellular mobile telephone service solved the problem of congestion, and its mass market acceptance made mobile service affordable to many customers.

Cellular mobile telephone service had lengthy delays in making its way to commercial introduction and availability. These delays were the result of the federal regulatory process. But once it was introduced, cellular service was and remains a phenomenal success.

The basic principles of mobile cellular telephone service were formulated at Bell Labs in the late 1940s. The technology to make the service economically feasible was not available, however, until the early and mid 1970s. AT&T had applied earlier to the Federal Communications Commission (FCC) for permission to offer an advanced mobile phone service (AMPS) based on the cellular principle, but the FCC wanted to determine how to introduce competition into the provision of cellular service. In early 1975, the FCC reallocated a portion of the UHF television band so that it could be used for cellular telephone service. The FCC opened 40 MHz of the 800-MHz radio band to any qualified common carrier, thus bringing competition to cellular service. In March 1977, the FCC granted authorization to the Illinois Bell Telephone Company to install and test a developmental version of AMPS in Chicago. The subsequent test was successful, and in 1983 the Chicago system offered the first commercial cellular service in the United States. The delays in offering cellular service were caused by the over ten years spent by the FCC in drafting regulation to stimulate competition. This regulation was finally approved in 1982.

Since the FCC had decided that it would be in the best interests of the public if competition were stimulated in cellular telephony, they issued two licenses in each area so that the public would have two service providers from which to choose. The wireline carriers were restricted to one license per area. As might be expected, there was an initial flood of applications to the FCC to obtain a license to provide cellular service. The FCC then encouraged competing applications to negotiate some form of consolidation. All this took more time and further delayed the introduction of cellular service, particularly for the nonwireline carriers.

Cellular service in each geographic area is provided by a wireline common carrier, which is a subsidiary of the local telephone company, and by a nonwireline common carrier. The radio spectrum space used for cellular service is divided in half to accommodate the two carriers serving a geographic area. The mobile equipment itself is available from a number of vendors. This equipment was initially costly, but these costs have dropped substantially as the service has become more widespread and used.

Basic principles

Prior to cellular telephony, mobile telephone service was obtained from a single high-power transmitter that served a single large geographic area. The service could serve only a small number of users and was very costly. With cellular mobile telephone service, a 50-MHz bandwidth in the 800-900 MHz radio band was used to create 832 two-way radio channels. This alone is a substantial increase over the few dozen channels previously available with conventional mobile telephony. However, the total number of customers served is increased further by reusing channels in the same geographic area. This feat is accomplished by the use of a number of low-power radio transmitters, each serving a small area, or cell, within the larger geographic service area.

The typical cell has a radius of about 6 to 12 miles. The low power of the transmitter means that the same channel can be used again in another part of the geographic area without causing interference. The entire geographic area is thus divided into cells, with each cell being served by its own transmitter. The cell configuration is chosen to minimize interference caused by the reuse of the channel—hence the use of the word "cellular." Usually, seven cells are taken to form a cluster, and the pattern of the cluster is repeated over and over again in the metropolitan area to be served. As shown in Figure 9.1, the cells can be depicted as hexagons; the actual shape of a cell varies with terrain and radio propagation. Cells serving a dense area with many users are usually allocated more channels. The channels are then reused in adjacent clusters. The spatial separation of cells using the same radio channels coupled with their low power reduces any cochannel interference to acceptable levels.

If congestion starts to appear, cells can be further subdivided, or split, into smaller cells using even lower powered transmitters. Thus, the system can grow gradually to serve more users as demand increases, although additional cell sites with antennas and network equipment are required. A larger number of users leads to lower prices for the mobile telephone instrument, and this leads to further increases in demand.

A key feature of cellular mobile telephone service is the mobility of the user. This mobility clearly extends across cells, and thus there is a need to track the user and to change radio channels for different cells.

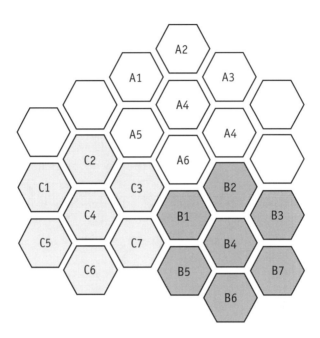

Figure 9.1   Cells are organized into clusters, with most cellular systems using seven cells per cluster. The clusters are then repeated over and over again to cover the whole geographic area served by the system. Although the cells are depicted as hexagons, their actual shape is quite irregular and depends on the terrain and radio propagation.

This dynamic tracking of the user along with changes in radio channel is called a handoff, and is a key feature of cellular service. The information necessary to cause a handoff is sent in the downlink speech channel, a technique called in-band signaling. Another essential feature is the use of a shared radio channel that is used to broadcast the numbers of the mobile units being called—a feature called paging.

To summarize, the basic features of wireless cellular telephone service are as follows:

- Low-power transmission;

- Frequency reuse through clusters of cells;

- Handoff from one frequency to another;

- In-band signaling; and

- Paging.

All the specific implementations of the various terrestrial mobile cellular systems around the world are based on these basic features and principles.

Multipaths

Cellular telephony uses high-frequency radio transmission, which is particularly susceptible to reflections that lead to multiple paths from the transmitter to the receiving antenna. Some of these multiple signals arrive out of phase with respect to each other and accordingly cancel. These cancellations cause fast fading of the received signal as the automobile is traveling down the road. Signal processing can be used to minimize the subjective annoyance of this fading by filling in missing portions of the speech signal.

One possible solution to the fading problem is the use of two receiving antennas at the mobile unit. The two antennas are separated by about half a foot. If the signal at one antenna is in a deep fade, then there is a strong probability that the signal at the other antenna is strong. An electronic switch compares the two signals and chooses the stronger. In this way, some of the fading problems caused by multiple-path (multipath) reflections can be corrected. This is known as switched space diversity and was used at one time by at least one manufacturer of mobile units for the Advanced Mobile Phone Service (AMPS) system in the United States.

Multipaths can be reduced by replacing omnidirectional antennas at the base stations with directional antennas that concentrate the radio signal in a narrow beam, thereby reducing reflections. The focusing of the antenna directional pattern to create pie-shaped sectors within a cell is called sectorization. Typically, there are three or six such sectors per cell.

The effects of fast fading can also be reduced by sending the radio signal in a number of frequency channels. This is accomplished by hopping across all the channels in a known repetitive pattern. In this way, if one channel is in a fade, the others most likely are not and most of the signal will get through. Frequency hopping is used in the Global System for Mobile (GSM) communication.

In newer digital cellular mobile systems, adaptive equalization of the received radio signal is used to combat multipaths. In effect, the reflected signals are subtracted from the received signal through the use of digital filters that dynamically change their characteristics in response to differing situations.

Implementation differences

All cellular mobile systems are based on the features and principles listed previously. The differences in various implementations are how individual radio channels are shared by users, how the speech signal is encoded, which radio bands are used, and how the radio carrier is modulated.

Advanced Mobile Phone Service (AMPS) was the first system to be implemented, and it was based on the technology of the 1980s. A baseband, analog speech signal is transmitted directly. Newer cellular systems encode the speech signal in a digital form. Digital encoding increases the bandwidth needed to sent the encoded signal, but bandwidth is severely constrained in cellular systems. Hence, some form of compression is usually used to encode the digital signal. Voice coders—called vocoders—have been investigated for decades, and the encoding used in cellular systems draws heavily on the results of this research. Most of the encoding schemes are a variant of a dynamic digital filter with varying coefficients whose values are obtained from linear predictive coding (LPC)—a technique invented at Bell Labs by B. S. Atal in the 1960s.

The frequency bands used for cellular were initially centered around 900 MHz. Newer systems operate as high as 1,900 MHz. The bands are divided into a number of radio channels with differing widths, depending on the implementation. The original AMPS system uses radio channels that are 30 kHz wide. Some newer digital systems have radio channels that are 200 kHz wide. The radio carrier for each channel is modulated through a variety of methods. Broadband frequency modulation is used for the radiofrequency modulation in the AMPS system. This technique was used to increase noise immunity for the low-power radio signal. Forms of phase-shift keying are used in some of the digital systems. Since all cellular systems use a number of separate radio channels, they all are based on a form of frequency-division multiple access (FDMA) at the radio-frequency level.

Multiple access

In the older AMPS system, a single voice signal occupies each entire 30-kHz radio channel. The newer systems attempt to increase their overall capacity by sharing a radio channel among a number of voice signals. This is accomplished through either time-division multiple access (TDMA) or code-division multiple access (CDMA), or (more usually) a combination of both.

The term "multiplexing" refers to the combining together of a number of separate signals to share a communication medium. This sharing through multiplexing is also referred to as multiple access and is usually accomplished through frequency-division and time-division multiplexing. In multiple access terminology, these are called frequency-division multiple access (FDMA) and time-division multiple access (TDMA). They are illustrated in Figure 9.2.

In frequency-division multiple access (FDMA), each signal corresponding to a telephone conversation is assigned its own exclusive band of frequencies for the duration of the call. In time-division multiple access (TDMA), the same band of frequencies is shared by all the calls, with a short burst of digital data from each conversation being sent alternately in a repetitive pattern. TDMA is usually performed in a number of separate bands—in effect, a combination of FDMA and TDMA.

A new method for multiple access that is being used in some newer digital wireless systems is code-division multiple access (CDMA). With CDMA, each digital signal is assigned a unique multiplicative code, which has the effect of spreading the spectrum to cover the entire assigned band. All the spectrum-spread signals are then sent on top of each other simultaneously and are separated at the receiver by an inverse application of the unique codes. Another form of spread-spectrum technology is frequency hopping, in which short bursts of data from each signal hop from one frequency to another in a known pattern.

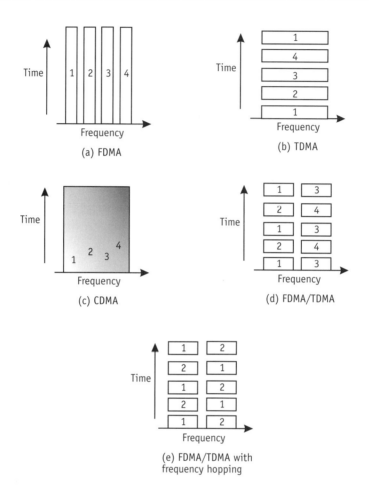

Figure 9.2 Various means are used to facilitate multiple access to a communication medium: (a) in frequency-division multiple access (FDMA), each signal corresponding to a telephone conversation is assigned its own exclusive band of frequencies for the duration of the call; (b) in time-division multiple access (TDMA), the same band of frequencies is shared by all the calls, with a short burst of digital data from each conversation being sent alternately in a repetitive pattern; (c) in code-division multiple access (CDMA), each digital signal is assigned a unique multiplicative code, which has the effect of spreading the spectrum to cover the entire assigned band—all the spectrum-spread signals are then sent one top of each other simultaneously and are separated at the receiver; (d) TDMA is usually performed in a number of separate bands, in effect, a combination of FDMA and TDMA; (e) with frequency hopping, short bursts of data hop from one frequency to another in a known pattern.

System operation

Cellular mobile telephone service is a system involving the equipment located at the mobile unit, the radio equipment located at the cell site, and a central switching office that controls the operation of the whole system and that interfaces to the public switched telephone network.

Each cell is served from its own master radio transmitter, receiver, and antenna. The antenna can be either omnidirectional or directional. An omnidirectional antenna would be situated in the center of the cell, and a directional antenna with a coverage of 120 deg could be situated at one of the vertices of the cell. A directional antenna is somewhat less susceptible to interference from other cells. Directional antennas are used at the center of a cell to create sectors, with three and six sectors being typical where such sectorization is implemented, as shown in Figure 9.3. The radio beams used with sectorization are more narrow, and hence are more powerfully concentrated and also are less subject to multipath reflection. The antenna and other equipment located at each cell site constitute a base station (see Figure 9.4). How signals are encoded,

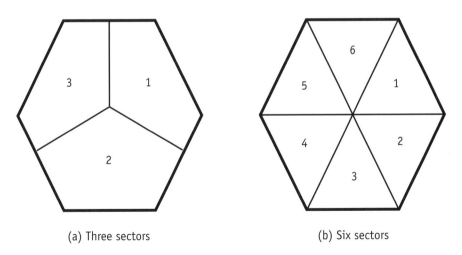

(a) Three sectors   (b) Six sectors

Figure 9.3   With sectorization, a directional antenna is beamed to divide a cell coverage into sectors, like wedges in a pie. Three (a) and six (b) sectors are typical. The narrow beam of the radio signal minimizes multipaths and can be more powerfully concentrated, thereby extending its range.

Figure 9.4(a)   Photograph of cell site, or base station. (*Source*: Bell Atlantic Mobile Systems ©1986. All rights reserved.)

combined, and transmitted determines the differences between the various implementations of cellular mobile telephone service.

Advanced Mobile Phone System (AMPS)

The very first implementation of the cellular mobile concept was the Advanced Mobile Phone System, or AMPS. This is an analog system in which each user fully occupies each radio channel of 30 kHz. AMPS is old technology in an industry that is rushing ahead in terms of market saturation and of advances in technology. But AMPS fully utilizes all the features

Figure 9.4 (b) Photograph of Figure 9.4a's cellular antenna. (*Source*: Bell Atlantic Mobile Systems ©1986. All rights reserved.)

and principles of cellular mobile telephone service previously described and listed. AMPS therefore is a good place to start our study of the workings of specific implementations of cellular mobile telephone service.

Each base station in AMPS operates in the 800-900 MHz range. A radio frequency band from 824 to 849 MHz is used to receive signals from the mobile units (the uplink), and a band from 869 to 894 MHz is used to transmit signals to the mobile units (the downlink). The wireline carriers use the so-called B band, and the nonwireline carriers use the A band; each carrier has 416 two-way radio channels. The 416 two-way radio channels for each system are divided over the seven cells that form a cluster, giving an average of about 60 channels per cell.

The base station transmits with a power of up to 100W, and a non-portable mobile unit transmits at a maximum effective radiated power of about 7W. To help control interference, the transmitted power of the mobile unit can be remotely decreased in steps. This is accomplished by a special code that is transmitted to the mobile unit and causes the mobile unit to decrease, or attenuate, its transmitted power to any one of eight prespecified levels. Each level decreases by 4 dB from the preceding level. The code that accomplishes this decrease in transmitted power is called the mobile attenuation code.

Because noise immunity is very important, frequency modulation (FM) is used for the radio transmission between the mobile units and the base stations. FM with a peak deviation of 12 kHz of the carrier is employed with an rms frequency deviation of 2 kHz for a normal speaker. The radio channels are spaced every 30 kHz in the allocated bands. Since the bands have a total width of 50 MHz, the maximum number of two-way radio channels is 832. Half of these channels go to the A band and the other half to the B band.

The base stations are connected by landlines to a central place called the mobile telephone switching office (MTSO), as shown in Figure 9.5. The MTSO is connected by trunks to the public switched telephone network. An important feature with cellular service is the ability to make telephone calls to the mobile unit over the switched public network as if the mobile unit was a normal telephone with its own 10-digit telephone number. Hence, the MTSO is connected to the public switched network via these trunks.

The mobile unit can move from cell to cell, and when this occurs, it will need to change the frequencies of the radio channels used to transmit to and receive from the base stations. This is known as a handoff. The relative received signal strength of the mobile unit must be monitored, and when it leaves the range of one base station and enters the range of another base station, information to specify the new frequencies must be sent to it. This information is transmitted as a short burst of data over the voice channel that is being used. The mobile unit needs to know when an incoming call is occurring and also must be able to dial out to set up a call. A shared data channel (also called a paging channel, a setup channel, or a control channel) is used for these purposes. The MTSO supervises and controls the entire operation of the system, including the assignment of

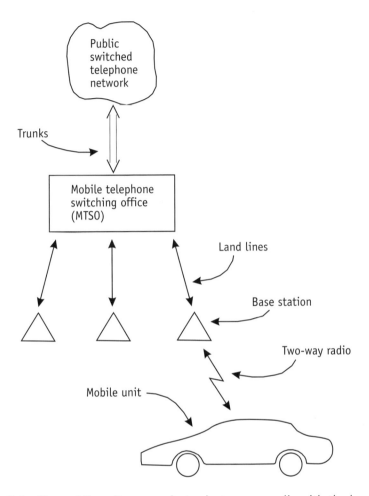

Figure 9.5 The mobile unit communicates by two-way radio with the base station. The base stations all communicate over landlines with the mobile telephone switching office (MTSO), which then maintains communication with the public switched telephone network (PSTN).

channels and the changing of frequencies during a handoff. The MTSO sends data over the landlines to the base stations for these control and supervisory functions.

The shared data channel transmits data at a 10-kbps rate. However, data encoding and message repeating reduce the actual information rate to 1.2 kbps, which is equivalent to a practical limit of 25 messages per second when fully loaded. Frequency shift keying is used, with a frequency deviation 8 kHz above and below the carrier frequency. The binary data is encoded using a biphase Manchester format, in which a 0/1 transition signifies a binary 1 and a 1/0 transition signifies a binary 0. The shared data channel is used to request service and to identify the mobile unit when its power is turned on, to assign the initial frequencies to be used for transmission and reception, and to terminate service.

A 34-bit binary number derived from the 10-digit telephone number assigned to the subscriber uniquely identifies each mobile unit and is entered and stored in the mobile unit. This number is sent by the mobile unit to the MTSO, where it is used for billing purposes. A 15-bit code entered into the mobile unit identifies the home system from which service is obtained. A 32-bit electronic serial number (ESN) is factory set into each mobile unit and can be accessed from the MTSO for security purposes.

The 50-MHz-wide radio spectrum available for cellular service is divided into two 25-MHz spaces that are assigned for use by the two separate mobile system operators in each area, a so-called A system and B system, as shown in Figure 9.6. The B system is operated by a subsidiary of the local wireline telephone company (a Bell company or an independent non-Bell company, such as GTE), and the A system is operated by a nonwireline carrier. The A system uses radio frequencies from 824 MHz to 835 MHz and from 845 to 846.5 MHz for transmissions from the mobile units and frequencies from 869 MHz to 880 MHz and from 890 to 891.5 MHz for transmissions to the mobile units. The B system uses 835 to 845 MHz and 846.5 to 849 MHz for mobile transmissions and 880 to 890 MHz and 891.5 to 894 MHz for base transmissions. Transmission from the base station to the mobile unit is called the forward direction, and transmission form the mobile unit to the base station is called the reverse direction. Each of the two systems has a total of 416 two-way radio channels available for its use in these frequency spaces, divided across the seven cells in a cluster.

Twenty-one channels in the A system, and 21 channels in the B system are used as shared data channels, or setup channels. The 21 setup channels are allocated across the cells, with usually one channel per cell. The mobile unit is programmed to scan all the setup channels and chooses

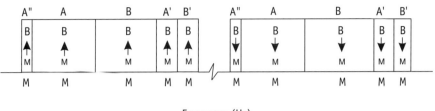

Frequency (Hz)

Figure 9.6 Two radio-frequency bands are used for cellular telephone service in the 900-MHz bands by the AMPS system. The band from 824 MHz to 849 MHz is used for transmission from the mobile unit to the base station (the uplink) and is shared by the A and B systems. The band from 869 MHz to 894 MHz is used for transmission from the base station to the mobile unit (the downlink) and is likewise shared by the A and B systems.

the setup channel with the strongest signal. This channel usually corresponds to the nearest base station. The mobile unit can be programmed to scan the setup channels for only one system or to scan both but with a preference for one system over the other.

A transfer from one radio channel to another is called a handoff. The MTSO makes a decision about whether a handoff is required by measuring the strength of the signal received from the mobile unit every few seconds. This periodic examination of the mobile's signal is called locating, which refers not to a precise geographic location but rather to signal strength. Thus, a mobile unit could be geographically closer to one base station, but because of terrain, its signal might actually be more strongly received at a more distant base station.

Supervision of a call is accomplished by out-of-band tones on the voice channel. A supervisory audio tone (SAT) of either 5,970; 6,000; or 6,030 Hz is regularly sent every ¼ second by the base station to the mobile unit, which then retransmits the tone back to the base station. If the tone is not received back at the base station for a long enough time, it is assumed that the mobile unit has ceased transmitting. The base station sends information (called the SAT color code, or SCC) to the mobile unit specifying the specific SAT to be used, and if the actual received SAT disagrees, voice communication is suspended. The round-trip delay of the SAT was at one time suggested as a gross measure of the physical distance of the mobile unit from the base station, but received signal strength is now used to determine handoffs.

A signaling tone (ST) of 10 kHz is sent from the mobile unit to the base station. The ST is used to acknowledge any orders from the MTSO, to perform flash requests comparable to the hook-flash of a conventional telephone, and to signal a release request when the user wishes to disconnect the call.

If the channel needs to be changed during a call, information is sent from the base station to the mobile unit over the voice channel as a very brief burst of digital data. The mobile unit decodes this information, changes its transmitting and receiving frequencies to the new channels, and establishes the voice circuit on the new cell. All this happens in a fraction of a second and is not noticed by the users. The AMPS cellular system is designed to handle calls arriving at a mean rate of one call per second in the densest cell.

The data channel is used by the mobile unit to request service. The identity of the mobile unit along with the telephone number of the called party are transmitted over this channel to the base station and then over landlines to the mobile telephone switching office (MTSO). The MTSO then establishes a conventional telephone circuit to the called party. When the mobile unit is turned on but is not being used for an actual call, it continuously monitors the strongest data channel for any transmitted paging messages containing its identification number. If the mobile unit recognizes its identification number in any of these paging messages, it responds to the page by sending an appropriate data message back to the base station over the data channel. It then receives information over the data channel telling it what specific voice channel to use to establish voice communication. Once the voice channel has been established, the base station sends a data signal over the voice channel to activate ringing at the mobile unit. The 10-kHz signaling tone is transmitted over the voice channel by the mobile unit to indicate that it is ringing. When this tone ceases, the base station knows that the mobile unit has answered, and the two parties are connected to begin their voice conversation.

Newer cellular systems

By 1990, cellular mobile telephony had been so successful that the demand had outstripped capacity. Hence, solutions were sought to increase the capacity of the existing allocation of radio spectrum space and also to use additional spectrum space. The AMPS system allocates a single user per 30-kHz radio channel. Digital encoding coupled with TDMA allows three users in the identical 30-kHz radio channel—the digital AMPS (DAMPS) approach, also known as IS-54. The European global system for mobile (GSM) communication is also digital TDMA-based, but uses TDMA within radio channels that are 200 kHz wide. An even newer approach is to spread the spectrum of the signals over a broad bandwidth and then send many signals at the same time using unique codes to separate them upon reception—the CDMA approach, also known as IS-95.

In these newer cellular systems, the stationary base station (BS) antenna site is known as a base transceiver station (BTS). A number of BTSs are controlled by a base station controller (BCS) and together form a base station subsystem (BSS), as shown in Figure 9.7. A number of BCSs then connect to the public-switched telephony network (PSTN) through a gateway mobile switching center (GMSC). The GMSC has access to a variety of databases containing such information as the equipment identity register (EIR), a home location register (HLR), and a visitors location register (VLR). The EIR contains the serial numbers of authorized mobile units and thus acts as protection against stolen mobile units. The paging channel is called a broadcast channel (BCH), and the actual two-way communication occurs over a traffic channel (TCH).

Global System For Mobile (GSM) Communication

The Global System for Mobile (GSM) communication was introduced in 1992 as a European standard and has achieved much success there. A variant of it is available in the United States. GSM initially stood for Gropue Spéciale Mobile.

GSM as first introduced in Europe operates in the 900 MHz band (called GSM900), and a newer system operates in the 1,800 MHz band

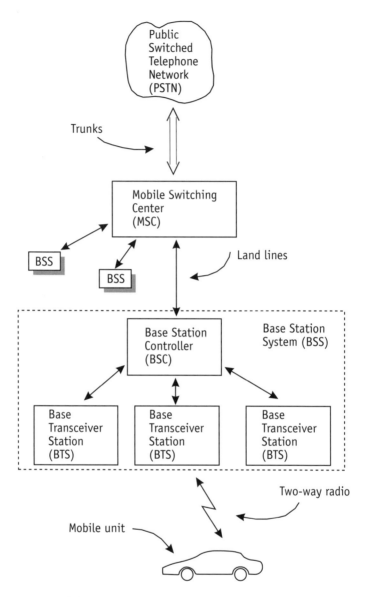

Figure 9.7   A mobile unit communicates by two-way radio with a base antenna at the base transceiver station (BTS). A number of BTSs are monitored and controlled by a BSC. The BSC and its BTSs form a base station subsytem (BSS). A number of BSSs are controlled and monitored by the mobile switching center (MSC), which also is the gateway to the public switched telephone network (PSTN). The MSC has access to various databases that contain information about the mobile users.

(called GSM1800). A GSM1900 system, intended for the United States, operates in the 1,900 MHz band. Other than operating in different radio bands, all the GSM systems are the same in terms of their principles of operation. An advantage of the GSM approach is that mobile phones can be manufactured that will work in all three bands and thus can be used in any country that has GSM.

In GSM900, the uplink radio connection is in a band from 890 to 915 MHz, and the downlink radio connection is in a band from 935 to 960 MHz. The uplink for GSM1800 operates from 1,710 to 1,785 MHz, and the downlink operates from 1,805 to 1,880 MHz. An uplink and a downlink taken together create a single two-way radio circuit, and such a pair is always chosen to be 45 MHz apart across the two links in GSM900 and 95 MHz apart in GSM1800. A radio channel within each band is 200 kHz wide. After allowing for a 200-kHz guardband, a total of 124 two-way radio circuits are formed in the uplink and downlink bands and are assigned across a cluster of seven cells in GSM900. GSM1800 uses 374 two-way radio circuits in its bands.

Eight users share each 200-kHz radio channel through the use of TDMA. This means that each user effectively has 200-kHz/8 = 25 kHz, which is comparable to the bandwidth assigned each user of AMPS. The GSM speech signals are encoded at 13 kbps using a form of predictive coding for compression, called regular pulse-excited, long-term prediction (RPE-LTP) and also known as residual-excited linear predication (RELP). The speech coder models the excitation source and the vocal tract, and encodes the information for every 20-msec segment of the speech signal into 260 bits. Thus there are 260 bits per 20 msec, or an overall data rate for the encoded speech signal of 13 kbps. The gross bit rate for the encoded speech signal plus channel coding for synchronization and control increases to 22.8 kbps.

Error correction is introduced along with interleaving of the signal so that any bursts of data that are destroyed can be reconstructed. These techniques are so powerful that two bursts lost out of eight bursts can be reconstructed perfectly. However, these error correction techniques add to the overall bit rate, which then becomes 33.85 kbps per speech circuit. The 200-kHz radio channel supports a data rate of 270.8 kbps for the eight circuits that are time-division multiplexed together. All the data for each circuit, including the encoded speech, synchronization, and error correction, is sent in a short burst of data lasting about 0.540 msec. Actually, time is allocated to ramp-up and than ramp-down gradually before transmitting the actual data. The time for this entire sequence of ramp-up, data, and ramp-down is 0.577 msec. Eight bursts, or time slots, of data taken together are called a frame. The radio carrier is modulated by a technique called Gaussian minimum shift keying (GMSK), which is a form of binary frequency shift keying (FSK).

Paging, signaling, and synchronization information are sent as bursts of data in a common control channel (CCCH), which is a logical channel created from the assignment of specific time slots in the flow of TDMA data in a radio channel. For example, the first time slot in the sequence of eight—a frame—might be assigned as the common control channel. This approach of creating a virtual channel within the stream of time slots is also used to create the broadcast channel (BCH) for paging of mobile units. When the mobile unit is first turned on, it begins an acquisition process to find the radio channel being used to send paging and control information. The process searches for the radio channel with the highest power and then examines that channel to determine whether it contains a peak in energy about 67 Hz above the center frequency of the radio channel. This peak corresponds to a sine wave that is transmitted in the control channel for frequency correction purposes—it is called a frequency-correction channel (FCCH). If the peak is not found, the mobile then searches other radio channels until it is found. The mobile is passive throughout this acquisition process, which typically takes only a few seconds.

Frequency hopping at the relatively slow frame rate is used—called slow frequency hopping (SFH). Each frame is sent in a repetitive pattern hopping from one frequency to another through all the available radio channels. Frequency hopping reduces the effects of fading. This is because fading is frequency dependent, and if fading is occurring at one frequency, it most likely will not be occurring at the other frequencies, and only a small amount of data will be lost.

The mobile unit is programmed to monitor the strength and quality of a sample of the received radio signals from adjacent cells. A report is constantly generated and transmitted back to the base station, which then determines when to switch the call to a new radio channel. The approach is called a mobile assisted handoff, or MAHO. A GSM handoff is decided on the use of radio signals received at the mobile unit as opposed to AMPS in which a handoff is decided on the strength and quality of the radio signal received at the base station.

A unique identity number identifies each mobile unit, and the mobile unit must have a card inserted into it containing this number before it can be activated. This card—called a subscriber identity module (SIM)—acts as an identity card for the mobile unit and affords some protection against loss of the mobile unit. The identity number is encrypted when transmitted over a radio channel. Signaling information and user information is encrypted to protect it from being intercepted and interpreted during radio transmission. As further security, each mobile user is assigned a temporary mobile subscriber identity (TMSI) for the time that the subscriber is using the mobile service.

With GSM, the system always knows the status of each mobile unit—for example, whether the mobile unit is on but idle or turned off. The mobile sends a message to the system when it moves to a new location so that the system always knows the location of each mobile unit—a process known as registration.

Digital AMPS (DAMPS)

Another solution to the capacity problem is the use of digital technology and TDMA to increase the capacity while staying within the present spectrum space. A digital approach can increase the capacity of each channel used, and the signal is more secure from someone listening in to the channel. With the old analog AMPS cellular system, anyone with a radio receiver tuned to the frequency of a cellular channel could listen to the conversation.

Digital AMPS uses a digital bandwidth compression technique called linear predictive coding (LPC). Three voice signals are time-division multiplexed together into a single radio channel. Each voice signal is encoded at 13 kbps using a form of LPC, and additional data is added for handshaking, noise immunity, and synchronization purposes. A length of 40 msec of each speech signal is analyzed and then sent in a burst. Three bursts corresponding to the three signals are sent sequentially every 40 msec within a 30-kHz cellular channel. The three signals time-division multiplexed together require a total of 48.6 kbps, which can be sent over a single 30-kHz radio channel. Each speech signal occupies its own data burst in the TDMA sequence. The final result is a tripling in capacity compared to conventional analog AMPS.

Digital AMPS operates in the same 30-kHz radio channels used by AMPS. In this way, an AMPS system can be gradually converted to DAMPS by assigning more and more radio channels to it. The radio signal is modulated using a form of differential phase-shift keying (DPSK). Modern cell phones can operate in either an AMPS or a DAMPS format. Such cell phones are called dual mode. They attempt to connect first using DAMPS and then if no capacity is available, switch to AMPS.

Code-Division Multiple Access (CDMA)

Another technology that can increase the capacity of cellular telephony is spread-spectrum transmission. With spread-spectrum transmission, each voice signal is sent over the full bandwidth of the total spectrum space. The trick is how to separate the jumble of received signals to concentrate on the desired single signal. One way is to digitize the voice signals and add a unique signature code to each by multiplying the signal bit stream by a unique signature bit stream. The receiver then searches for the unique signature code for the desired signal. This is somewhat similar to way we can listen to one conversation at a cocktail party even though our ears are being bombarded by dozens of speech signals. Since each multiplexed signal has its own unique code, the technique is called CDMA.

Another spread-spectrum technique is frequency hopping, in which each signal continuously switches channels in a known pattern. Again, the receiver searches for the appropriate pattern for the desired signal. Spread-spectrum transmission can be less prone to interference and thus allows more signals to share the same spectrum space. Slow frequency hopping is used in the GSM system.

Spread-spectrum technology in the form of CDMA was advocated and developed for cellular mobile telephone service by Qualcomm,

Incorporated. Qualcomm claims an increase of from 10- to 20-fold in capacity over conventional analog AMPS. The spread-spectrum system was known initially as Qualcomm CDMA and then became formalized for North America as interim standard IS-95, adopted by the Cellular Telephone Industry Association (CTIA). It is now also known as CDMA-One. It operates in the same 900-MHz band as AMPS. Each radio channel has a width of 1,250 kHz for a total of 20 radio channels in each direction. The channels operate in uplink and downlink pairs separated by 80 MHz. CDMA is also being used by many of the new PCS systems in the United States operating in the 1,900-MHz band.

The speech signal is encoded and compressed through a technique called Qualcomm code-excited linear prediction (QCELP) at a bit rate that can vary adaptively from 1 kbps to 8 kbps. The speech data along with error-correction coding results in a gross bit rate that varies from 2.4 kbps to 19.2 kbps. In practice, the speech quality of the lower bit rates is usually not acceptable, and the higher rates must be used. The bit stream is multiplied by a pseudorandom code at a much faster bit rate. This has the effect of spreading the spectrum over a much greater bandwidth; the code is called a spreading code. The spread signal fully occupies the 1,250-kHz radio channel.

A number of signals are all sent on top of each other in the same 1,250-kHz radio channel. At the receiver, the appropriate pseudorandom spreading code is applied to extract the desired signal. All unwanted signals appear as random noise and are ignored.

Qualcomm claims considerable superiority of CDMA over all other methods for cellular mobile telephone service in terms of its capacity. But CDMA only works if all the received radio signals are about the same strength. If one signal is much larger than all the others, it will dominate and the others cannot be recovered. This means that mobile units have to adjust their transmitting power so that the signals received at the base station are all the same power regardless of distance. There are also quite complex timing issues with the received signals. Complexity is not necessarily a problem with today's technology, but there are far simpler wireless systems already available in the marketplace. Whether CDMA will achieve its great promise remains unclear at the time of the writing of this book.

Intersystem roaming

A cellular mobile unit will work anywhere there is cellular service, provided that an agreement exists to hand billing back to the unit's home system. A mobile unit could roam all over the country and make telephone calls. Billing is not a problem since the mobile unit contains its own identification number along with an identification of its home system. Reaching the mobile unit is another story, however.

One way to call a roaming mobile unit is for the caller to know the area where the roamer is located and to dial a roamer access number for that area. The caller then enters the 10-digit number for the mobile, and a page is made by that local system. The problem with this method is that the caller must know the roamer access number for the area where the roamer has roamed. A better way is GTE's Follow Me Roaming® service. Upon entering a foreign cellular area, the roamer sends a special code to the cellular system. The foreign cellular system then notifies the roamer's home system about where the roamer is located. Any calls to the roamer are then automatically forwarded to the foreign cellular system where a page is made to reach the roamer to complete the call.

The continuing search for increased capacity—a personal assessment

Cellular telephone service has almost been too successful, and some systems are running out of capacity. Cells can only be split a certain number of times before interference becomes a serious problem. The question has now become how to achieve more capacity.

The clearest and simplest solution is to add more spectrum space for cellular radio transmission. However, that space would need to be taken from spectrum space presently allocated to UHF television transmission, but the UHF broadcasters do not want to relinquish any space. Furthermore, some of the systems proposed for high-definition television (HDTV) would use the UHF spectrum. The TV broadcast industry and the cellular industry are engaged in a battle over this issue of spectrum space. My view is that with cable television passing nearly 90 percent of U.S. households, along with the strong penetration of VCRs, there is more than enough television diversity and that some more UHF space should be made available for two-way cellular service.

In the end, the great debate between CDMA and other methods over which system has the most capacity must accept that fact that Claude E. Shannon showed in 1948 that the maximum signal carrying capacity of a communication channel is finite and cannot be exceeded. Comparisons must be made on an equal basis to be assured that one is really comparing like items. As an example, the CDMA system takes into account the fact that telephone conversations are really one-way most of the time in the sense that one party is speaking while the other is listening. We saw how this can be used in transoceanic cable systems to more than double capacity using TASI equipment. Although CDMA uses this kind of an approach to increase its capacity, other cellular systems could too, but do not.

A simple way to increase the capacity of a cellular system is the use of single-sideband amplitude modulation (AM) transmission. However, noise and interference can be a problem with amplitude modulation. The advantage of frequency modulation (FM) is that it has a capture effect in that the strongest FM signal is locked onto by the receiver, and all weaker interfering signals are ignored. Although single-sideband amplitude modulation is very efficient in its use of spectrum, amplitude modulation is seriously degraded by the fading that is encountered in cellular wireless transmission. Advances in analog technology, perhaps coupled with digital processing, may someday in the future lead to innovations that could make amplitude modulation practical. The increase in capacity that could be possible with analog technology would be most impressive.

Consider the AMPS system with its 30-kHz wide radio channel allocated to only a single user—a tremendous waste of valuable spectrum space. Each baseband speech signal requires 4 kHz. If single-sideband suppressed-carrier modulation were used in a simple FDMA approach, the capacity could be increased sevenfold. If the one-way nature were used, the capacity could be doubled to a 14-fold increase. From my years at Bell Labs in the 1960s, I recall analog vocoders operating at a ten-to-one compression that sounded perfectly natural. If they were used, the overall capacity could be yet increased by a factor of ten—a net increase of 140-fold! Clearly, we have not yet seen the end of the progression of innovation in cellular technology.

It is quite conceivable, however, that ways will be invented to improve upon conventional FM and AM techniques to increase the capacity of a cellular system without creating serious interference problems. Whatever happens to increase capacity, the problem of standardization looms ahead. The present analog FM scheme is now established as the standard for millions of cellular users. The challenge for the cellular industry is to choose a new standard and then migrate toward it while still supplying service to all the users of the older technology. This is no easy task!

Satellite mobile communications

With terrestrial-based mobile communications, the base stations are fixed and the users are mobile. A number of fixed base stations are required because of the low-power and frequency nature of cellular wireless telephone service. One alternative approach is to place a single base station high in the sky above the region to be served. As we saw in a previous chapter, communication satellites accomplish this, but if placed in geostationary orbit are so high that the round-trip delay is a problem for two-way communication. A solution is to place the satellites in a much lower orbit. Low earth orbit (LEO) satellites have already been launched to offer mobile communication.

The first LEO system was developed by Motorola and is called Iridium. It uses 66 satellites in circular polar orbits about the earth in six orbital planes at a height of 420 miles (780 km). Each orbital plane contains 11 satellites so that there will always be a satellite above the area to be served. The satellites move across the sky while overhead, and a new satellite must come into place before the current one disappears over the horizon. Each satellite is in use for about six minutes. This means that handoffs must occur to maintain continuous communication. Each satellite in the Iridium system has a capacity of 3,840 full-duplex circuits with each circuit at 4.8 kbps. Compressed speech or computer data can be sent in a circuit. The Iridium system is scheduled to be operational in 1998, and the satellites have already been launched. At such a low orbit, the satellites encounter friction, will fall back to earth after about five years, and hence will need to be replaced quite frequently.

A better compromise with the need to avoid delay yet be at a height so that a reasonable number of satellites is required with fair lifetimes is to place the satellites in medium earth orbit (MEO). The TRW Odyssey system is a MEO system with satellites at a height of 5,575 miles (10,354 km). Twelve satellites are to be placed in three orbital polar planes. Each satellite will be overhead about two hours and will have a lifetime of 15 years. There will be a total of 2,300 full duplex circuits per satellite.

The Globalstar system being promoted by Loral, Qualcomm, and others will use 48 satellites in eight inclined circular orbital planes at a height of 754 miles (1,401 km). The satellites are expected to have lifetimes of five to seven years, will be overhead about 11 minutes, and can each serve 288 full-duplex circuits.

These LEO and MEO satellite systems are very costly and utilize technology that is yet untested. Since most of the surface of the earth is oceans, the satellites will be useless most of the time unless over large metropolitan areas. This means that their useful duty factor is very low. There are few people in rafts in oceans or in the rain forests of South America desiring telephone service and willing to pay a small fortune a minute. Yet a group of investors, including at least one billionaire, are supporting the Teledesic LEO system which plans to use 840 satellites in 21 polar planes at a development cost of $9 billion.

Satellites in geostationary earth orbit (GEO) can also be used for mobile communication if the user is willing to tolerate the round-trip delay. Mobile units for these GEO systems are the size of a laptop computer and cost a few thousand dollars. A call costs about $1.50 a minute. If there is any real market for personal satellite communication, the existing GEO systems already satisfy it.

It most certainly is far too early to tell which, if any, of these approaches will achieve commercial widespread viability. Perhaps one or more will discover very special "niche" markets that can afford the price and have few alternatives. But the obstacles to overcome are so many that I wonder whether the LEOs and MEOs will ever amount to anything other than a great deal of overpromotion and hot air!

One hot-air approach that could make sense is a hot-air balloon situated high above a metropolitan area. The balloon would carry a microwave transceiver system to offer mobile telephone service. The balloon would be at a low enough altitude to avoid any delay problems and also could be retrieved for servicing. Such hot-air balloon systems have been seriously suggested.

Cellular telephony—an assessment

Cellular telephony involves very sophisticated technology. The mobile unit is capable of being tuned automatically to any one of the two-way radio channels and can transmit and receive binary data and act on that data. The mobile telephone switching office must monitor thousands of calls, determine when handoffs are needed, send and receive data transmissions from the mobile units, and determine billing.

It is astonishing that mobile units have so shrunk in size that portable cellular phones that fit in a pocket are commonly available. The advances in cellular technology are most impressive, but are also the cause of ever-changing standards that appear to defeat compatibility. The advance of technology still has some challenges to overcome. One is how to locate physically a mobile unit so that the mobile can be handed to the appropriate public safety facility in the case of a call to 911. Appropriate localization methods are under investigation, but some users might be concerned about the potential loss of physical privacy.

There is one issue over cellular telephone service that could impede its use and ever-continuing growth. This issue is the health safety of the microwave radiation from the small antenna in the mobile unit. Mobiles transmit as much as 1 W of microwave power, and this radiation is directly against the side of the user's head. Although using a cell phone for a few minutes a day is probably not a hazard, I wonder about the hazard of daily use of an hour or more, as seems so common for some heavy users. The effect will certainly be cumulative, and will appear far in the future in the form of a unilateral brain tumor. Cellular telephone service is only 15 years old, but such cumulative health effects could take a few decades of use to appear. We thus do not yet know the answer to the potential health hazards of heavy use of cellular phones. We do know, however, that automobile accidents linked to cell phone use while driving are on the increase.

The continued success and growth of wireless communication has created a great demand for increases in capacity. But because the radio spectrum is limited, ways must be discovered to use the existing spectrum more efficiently. This had led to the use of speech compression to reduce the bandwidth of each speech signal so that more voice circuits can share the same channel. As we saw earlier in this chapter, such compression is accomplished at bit rates as lows as 8,000 bps rather than the usual 64,000-bps rate of uncompressed digital speech. Bit rates lower than about 8,000 bps are not acceptable to most consumers since the compressed speech sound rough and unnatural. Digital wireless implies to consumers that the quality will be similar to a CD. Alas, it is not. Compression is usually a compromise with quality.

Compression algorithms are designed for speech but do not work well for computer data, which requires as much digital capacity as possible. When used to access the Internet or send computer data, cell phones need to bypass the speech compression algorithms and offer direct access to the data stream. But these data streams—usually about 13 kbps—though acceptable for compressed speech communication, are low by today's standard of 56 kbps for modems that work on regular phone lines.

Yet, more important than all the technological sophistication is the ease of use of cellular telephony. When one is driving along a busy freeway, bumper to bumper, at 65 miles per hour, it must be easy to place and receive a cellular call. The dialing buttons must be well placed for ease of use, so calls can be placed without taking one's eyes off the road ahead. The challenge for future cellular phones for the car is the user interface. This is one area where synthetic speech could augment the visual display of a telephone number. In cellular dialing, the user enters the number into the mobile unit, and only after the complete number has been entered does the unit send the number to the base station. The user can review the dialed number on a visual display before sending it out, but a synthesized-speech "read-out" of the number would perhaps be most useful during driving. So, too, would be speech recognition so that the user could speak the digits of the number to be called, as is done in some newer cellular phones for cars.

Cellular mobile telephone service continues to be very exciting with great growth potential. Some futurists predict a day when everyone will carry their own personal telephone, served by some form of cellular service. In fact, the existing wireline telephone network could be bypassed through cellular phones. This had led some long-distance companies to investigate the use of a form of fixed wireless as a substitute for the copper wire of the local loop controlled by the local telephone companies. Thus far, the wireless local loop (WLL) seems too costly to compete with copper wire for such fixed applications.

With portable phones, car phones, cellular phones, and cordless phones, we have the ability to be reached and to reach anyone by telephone anywhere at any time. Whether we will all want to continue to be that easily reached at all times, though, remains to be seen.

Selected bibliography

Calhoun, G., *Digital Cellular Radio,* Norwood, MA: Artech House, 1988.

Falconer, D. D., F. Adachi, and B. Gudmundson, "Time Division Multiple Access Methods for Wireless Personal Communications," *IEEE Communications Magazine,* Jan. 1995, pp. 50–57.

Frieden, R., "Satellite-based Personal Communication Services," *Telecommunications,* Dec. 1993, pp. 25–28.

Harris, J., "GSM Basics: An Introduction," *Microwave Journal,* Oct. 1996, pp. 84–92.

Mangir, T. E., "Wireless via Satellite: Systems for Personal/Mobile Communication and Computation," *Applied Microwave & Wireless,* Jan./Feb. 1997, pp. 24–42.

Noll, A. M., "The extraterrestrials are coming," *Telecommunications Policy,* Vol. 20, No. 2, 1996, pp. 79–82.

Prentiss, S., *Introducing Cellular Communications,* Blue Ridge Summit, PA: Tab Books Inc., 1984.

Redl, S. H., M. K. Weber, and M. W. Oliphant, *An Introduction to GSM,* Norwood, MA: Artech House, Inc., 1995.

Webb, W., *Understanding Cellular Radio,* Norwood, MA: Artech House, Inc., 1998.

"Advanced Mobile Phone Service," *Bell System Technical Journal,* Vol. 58, No. 1, Jan. 1979.

10

Data Communication

Introduction

The world of telecommunication has its "two cultures"—namely, voice and data. Voice communication systems are designed by telephone engineers; data communication systems are designed by computer scientists. This has produced a schism between the circuit switching of telephony and the packet switching of data communication. In reality, each culture needs and can learn much from the other, for there are great synergies in applying the best of each approach in an overall integrated-network system solution.

Dial-up telephone network

Vast amounts of information are stored in computerized data banks, or databases, such as credit histories, airline schedules, and law cases. This information is accessed remotely using a personal computer by simply establishing a telephone connection to the database. Thus, the switched telephone network is the most ubiquitous way of communicating data. However, computers use binary signals—in the form of binary digits, called bits—for communication, but the telephone network has been designed for analog, speech-like signals. Hence, the binary signals required by the computer are converted to analog tones that can be conveyed over the telephone network.

The tones are carrier signals, and certain of their parameters are varied according to the binary data. The process of causing these variations is called modulation, and the process of extracting the binary data is called demodulation. A device is needed between the computer and the telephone line to perform the translation between the binary data required by the computer and the modulated tones required for transmission over the telephone network. Because communication is a two-way affair, this device must perform both modulation and demodulation of the tone, depending on the direction of the communication. Such a device is both a **mo**dulator and a **dem**odulator, and thus is called a **modem** for short. A data terminal usually has a modem built into it.

By way of a computer and a modem, a person can use the dial-up telephone network to reach most computerized databases. The speed of the communication can be as low as 300 bps to as high as 56,000 bps, with 28,800 bps and 33,600 bps being popular intermediate speeds. The 4-kHz bandwidth of a telephone circuit along with the signal-to-noise ratio limit the data capacity of the dial-up telephone network. Most modems allow simultaneous, two-way communication—a type of communication that is called full duplex. A type of communication that is one-way at a time and that requires a reversal of the direction of the communication to achieve two-way communication is called half-duplex. A strictly one-way type of communication, such as television broadcasting, is called simplex.

The dial-up telephone network is ubiquitous, easy to use, and very economical for much data traffic. However, it does have some problems for data communication. Longer local loops have loading coils, and these loading coils restrict the higher frequencies that can be carried over the loop, thereby restricting the upper limit on the data capacity. For certain types of high-speed data communication, the loading must be removed.

The echo suppressors that were used on long-distance circuits played havoc with full-duplex data communication and had to be removed for certain types of full-duplex data communication. The echo suppressors did not allow simultaneous two-way communication, and the time required for them to switch the directionality of the circuit was too lengthy for the high speeds of data signals. This problem is eliminated with modern echo cancelers.

The need for high-speed downloading of information from the Internet has created a demand for high-speed modems. Today's 56-kbps modem is at the limit for what is possible within a 4-kHz channel. This bandwidth restriction comes from the 4-kHz antialiasing filter at the analog-to-digital converter at the central office switch or at a digital voice multiplexer. The actual bandwidth of the copper wire of the local loop can easily carry a few megabits per second at loop lengths of less than 18,000 feet. This has led to the recent development of technology for digital subscriber lines, such as asymmetric digital subscriber line (ADSL), which is discussed later in this chapter.

ASCII

Most of the information sent as data over communication networks is text in the form of alphabetic and numeric characters. These characters are encoded as binary information using the American Standard Code for Information Interchange (ASCII). ASCII uses seven bits to represent a total of 128 possible characters, including uppercase and lowercase letters, numbers, special control symbols, and such characters as /?&*percent#. An extra eighth bit is added as an error-detection bit. It is set to either a 0 or a 1 to make the total number of 1s in the eight-bit string either an odd or an even number (called odd parity or even parity). ASCII is referred to as an eight-bit code. Figure 10.1 shows the ASCII transmission for the character "N."

Some data terminals transmit a single ASCII character at a time. The receiving equipment does not know when a transmission might occur, and hence the idle state of the transmission must be changed to alert the receiving equipment that a character is about to be sent. The idle state usually is a continuous transmission of 1s, called marks, so that any break

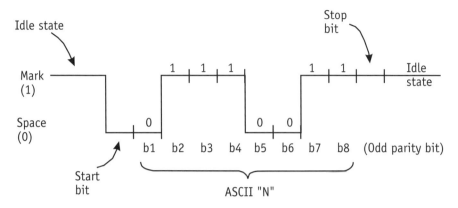

Figure 10.1 The character N is encoded in ASCII as a series of eight bits represented as marks and spaces. The signal arrives left to right, with the first arrival being a start bit to alert the receiving equipment to expect a character. Next come the seven bits for the ASCII character, with the eighth bit being an odd parity bit. A stop bit indicates the end of the transmission.

in the line can be ascertained. Thus, the first bit sent to alert the receiving equipment is a 0, called a space. This is followed by the eight bits for the ASCII character. The least significant bit b_1 arrives first, followed by the other seven bits, with the parity bit arriving last. Some time must be allowed for the circuit to return to the idle state, and hence a mark is sent at the end as a so-called stop bit.

This form of data transmission in which a single character is sent at a time is called asynchronous transmission. Asynchronous transmission is wasteful of capacity since a start bit and a stop bit must be sent for each character. A more efficient form of data transmission is synchronous transmission, in which a whole block of characters is sent in one long transmission. Some special codes start the transmission to synchronize the receiving equipment to the precise rate of transmission. This type of transmission is found in message and packet switching.

Message and packet switching

The telephone network is a line-switched, or circuit-switched, system. A complete, dedicated connection is first established, either physically or virtually, and then actual voice conversation begins. The physical or virtual connection is maintained for the full duration of the call, even if no one is speaking. This type of system is fine for speech communication, but it entails numerous problems for data communication.

Unlike speech communication, most applications of data communication consist of short bursts of information with a fair amount of dead time between each burst. The circuit-switched telephone network in which a complete connection is maintained for the duration of the call is very wasteful for this type of bursty data communication. Humans are willing to dial a telephone number and wait seconds for the connection to be established. Computers demand instant access and cannot wait idly for a circuit connection. Many computer users leave their computers connected to the network for long periods of time even though they are not sending any data. Short bursts, instant access, and long holding times characterize data communication.

Message switching and packet switching are appropriate for data communication. With message and packet switching, the intended destination is added to the data to be sent. The data along with its destination is gradually switched, stored, and forwarded through the data network to its final destination. This type of switching is also sometimes called store-and-forward communication.

In general, a message is one long burst of information. This long burst of information may be broken up into shorter bursts of fixed lengths, called packets. A data packet includes both the destination address and the source address. The actual data is typically about 1,000 bits. As shown in Figure 10.2, packets may be switched and transmitted over a number of different paths before reaching their final destination, where they are reassembled into the complete message. With message switching, shown in Figure 10.3, the message is switched and transmitted in its entirety over the data network. The message may be stored temporarily for some time along its transmission while waiting for an available circuit with sufficient capacity.

A packet-switched data network has nodes at which computers examine the packets and then send them along toward their final destination. Different paths may be taken by different packets, depending on the available transmission capacity between the nodes at the time the packet is transmitted. Digital circuits form the transmission capacity between

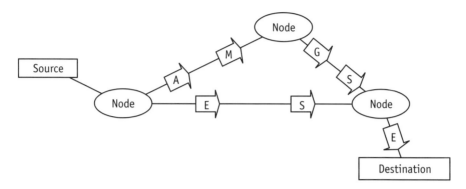

Figure 10.2   With packet switching, a complete message is broken into a series of shorter packets. The packets travel separately along the network from their source to the destination. The packets are assembled at the destination to form the original, complete message.

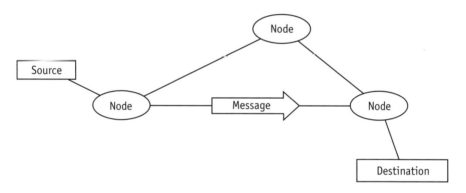

Figure 10.3   With message switching, a complete message travels as a whole along a data network from its source to the destination. The message may wait at nodes along the way for sufficient transmission capacity.

the nodes. Depending upon the traffic, the digital capacity could be as little as a single 64-kbps circuit, a 1.54-Mbps DS-1 circuit, or even higher.

Protocols and standards

When we speak over the telephone, certain agreed-upon rules and procedures govern our conduct. We answer the telephone by saying "hello."

We wait for a pause in the other person's conversation before interrupting. At the end of our conversation, we say "good-bye." Similarly, we require agreed-upon rules and procedures to ensure the orderly transfer of data between digital devices that are interconnected by communication facilities. These rules and procedures are called data communication protocols. The manner in which a terminal acknowledges that it is ready to receive data is an example of a protocol.

Many interfaces are encountered in data communication, such as the connection of the terminal to the network and how information displayed on the screen applies to a specific application. Clearly, interfaces and protocols must be standardized, or total chaos would result. To add some sense of order to the many interfaces, protocols, and standards encountered in data communication, the International Standards Organization (ISO) has created a conceptual model for data communication consisting of seven separate levels, or layers. Since the model is intended to facilitate the open interconnection of data systems, it is called the Open Systems Interconnection (OSI) model. The electrical connection between a modem and a terminal is an example of one of the interfaces considered by the OSI model for which standards are specified.

Although the seven-layered OSI model is intended to eliminate confusion, the seven layers and what they do still seem complex and confusing. The lowest layer (layer 1) is concerned with hardware and its physical and electrical connection. The highest layer (layer 7) is concerned with the specific application of the communication—for example, order entry or funds transfer. The seven layers are as follows:

1. Physical layer;

2. Link layer;

3. Network layer;

4. Transport layer;

5. Session layer;

6. Presentation layer; and

7. Application layer.

The first three layers (physical, link, and network) are concerned with routing of data from one piece of equipment to another. The last four layers are concerned with the actual dialogue for a particular application. The user interacts with the highest layer, the application layer. Information is then transferred from layer to layer and finally across the communication network to the receiving end, where the information ascends to the application layer at the host computer. Each layer packages the information differently, but the contents are unchanged. The "user" may be a computer program in some applications.

As mentioned, the three lower level layers apply to the method of data transport through the communication network. The physical layer is concerned with the physical and electrical interconnection of hardware. For example, the type of coding used to represent a bit by a change in voltage would be specified in the standards for the physical layer. The physical layer is concerned with the cable and wires used to interconnect data equipment, such as the RS-232-C standard for specifying the various voltage levels and signals on a 25-pin connector. The link layer is concerned with groups of bits, called frames. The use of checksums for error detection and correction is specified at the link layer, for example, along with protocols for acknowledging the receipt of data. The use and placement of the address for a data message sent over a packet-switched network is specified in the protocols for the link layer. The network layer is concerned with the actual routing and transmission of the data through the network. For a packet-switched network, the rules governing the actual routing of packets from node to node and to the final destination are specified in the standards for the link layer.

The four higher level layers are concerned with the specific application and its software. The application layer, for example, specifies the task or job that the user is accomplishing—for example, electronic mail or funds transfer. The presentation layer is concerned with such areas as the encryption of data, the positioning of the cursor on the display screen, the encoding of graphical symbols for display, and the standards needed for representing various types of data files. The time between when a user first enters a computer application ("logs in") and when the user finishes the application ("logs out") is called a session. The session layer is concerned with session-specific areas, such as user identification and recovery from transport failure. The transport layer is concerned with such areas as how data is handled when errors are detected and how to handle data overflows from network congestion.

Standards are available for most of the layers, but engineers continue to refine the definition of some of the intermediate layers and their appropriate standards and protocols. IBM has its own layered data networking protocol, called System Network Architecture (SNA), and Digital Equipment Corporation (DEC) has its own, called Digital Network Architecture (DNA). The implementation of DNA is called DECNET. One of the earliest packet-switched networks was ARPANET, developed by the United States Defense Department's Advanced Research Projects Agency. ARPANET was originally developed to allow large computer systems to share resources; the actual use was to send messages, or electronic mail, between the computer users. The ARPANET has evolved into today's Internet, which is described in the following section and uses the TCP/IP protocol. Other data network protocols are Xerox's Ethernet™ and Apple's AppleTalk™. SS7 is yet another data network with its own protocols.

Information for data transmission over specialized data networks is organized into a grouping called a frame. The beginning of a frame has a header containing address information and a trailer to indicate the end of the frame. A large block of characters forms the message to be transmitted, and the header and trailer are appended at the link layer. One popular protocol for this type of transmission is IBM's Binary Synchronous Communication, also known as BSC or as BISYNC. A newer IBM protocol is the Synchronous Data Link Control (SDLC), which accommodates any number of data bits and includes sophisticated error checking.

The use of many different protocols for different types of databases and computers can be a problem. One solution is to fix a place in the network where conversion from one protocol to another can occur. These protocol conversion places are called gateways, and they interconnect different data networks, databases, and computers.

The Internet

What is the Internet? At the broadest sense, it is a vast worldwide database of information and electronic communication that is accessed from personal computers. In the narrowest sense, it is a network of packet-switched data networks. Access to the Internet for many users is from a personal computer connected to a modem that uses the local telephone network to reach an Internet access provider. The Internet provider has a pool of modems to serve its users. The modems are multiplexed onto a high-speed stream of data connected to an Internet access point-of-presence (POP). The Internet is a great success because of its ease of use, made possible by a multimedia presentation of information with friendly links from one database to another. The World Wide Web is a term used to describe the Internet in the broadest sense.

The Internet utilizes packet switching. Packet switching was used in the late 1960s in the ARPANET, operated by the Advanced Research Projects Agency of the United States Defense Department. The ARPANET connected the large computer systems of the ARPA research partners and was intended primarily to facilitate the sharing of computer resources. Instead, it was used mostly to exchange documents and messages today's e-mail. The original ARPANET was formed from a single 56-kbps line leased from a common carrier.

In the early 1970s, the National Science Foundation (NSF) wanted access to the ARPANET for its university community. This became a policy issue since ARPA, as a federal agency, was not licensed to be a common carrier and hence could not sell access to others. The NSF then developed its own packet network—the NSFNET—which assumed the ARPANET. The backbone packet-switched network was called the National Backbone Service and was supported financially by the NSF. In effect, packet data communication had become nationalized in the United States—a move that was in direct contradiction to White House policies of the early 1970s. In 1985, the NSFNET connected supercomputer sites over a packet network operating at 56 kbps. In 1988, the capacity was increased to lines operating at 1.544 Mbps, and then in 1990 the capacity was upgraded to lines operating at 45 Mbps. IBM supplied the Internet Protocol routers, and MCI supplied the transmission lines. The NSF supplied the funding for the backbone network and a number of regional packet networks.

In 1995, because of the heavy commercial use of the packet network, now known as the Internet, the NSF ceased its support of the backbone and initiated a gradual cessation of its support of the regional networks.

However, once federal support begins, it rarely ever fully halts. In the mid 1990s, the National Science Foundation and MCI began the development of a very high performance Backbone Network Service (vBNS) utilizing ATM switches and IP routers. The vBNS was positioned as a research network for the academic community rather than the commercial venture that the Internet had become. The vBNS has been operational since April 1995. The backbone transmission for the vBNS was over SONET at the OC-12 level at 622.08 Mbps. In 1998, it consisted of 12 POPs at various MCI locations and four additional POPs at various supercomputer sites. The SONET backbone fiber spans 25,000 km. The vBNS is part of the MCI Hyperstream network offering ATM service commercially. A total of 120 universities in the United States are cooperating to interconnect using the vBNS and existing networks—the so-called Internet2 project. The former ARPA—now known as the Defense Advanced Research Projects Agency (DARPA)—has begun the development of the Next Generation Internet (NGI) as an initiative to deliver 100 times the capacity to 100 participants.

The protocol used for the Internet is the Transmission Control Protocol/Internet Protocol (TCP/IP). The TCP is at the transport layer, and the IP is at the network layer. TCP handles such functions as flow control, sequencing, reliability, status, and session closing. The data plus header information form a datagram. A datagram plus additional header information along with a trailer form a frame—a protocol data unit (PDU). An IP address consists of a network address and a host address, which together create a 32-bit address.

Packet networks

The ARPANET and its development into the Internet were the earliest packet switched networks. The technology has developed over the decades, along with new terminology, standards, and approaches.

A protocol data unit (PDU) is a length of information transmitted, or relayed from router to router, over a data network. For example, in the T1 system, a PDU consists of 24 bytes of digitized speech plus a single framing/synchronization bit. A byte, sometimes called an octet, is eight bits.

The X.25 standard is a computer-to-computer communication protocol for data sent over a packet-switched network. The X.25 standard is an interface standard. In X.25, a PDU consists of a packet plus a link control field. The packet consists of a data field of 128 bytes and a header. The link layer of the X.25 protocol assures that the packet is delivered error-free. Frame relay is a new protocol intended to replace the X.25 standard. The PDUs in frame relay are variable-sized.

Cell relay uses PDUs of fixed length, consisting of 48 bytes of data plus five bytes of header information. The header information includes a virtual path identifier (VPI), a virtual channel identifier (VCI), and header error control (HEC). Asynchronous Transfer Mode (ATM) is a cell relay system using fixed-length PDUs that are 53 bytes long. ATM is asynchronous, which means that precise timing is not needed. ATM has been promoted as the single architecture that can carry data and voice over the same packet-switched broadband network. Although data communication can accommodate short delays, voice communication is very intolerable to such delays. Furthermore, the header information in each cell is costly overhead for voice communication. There is not full acceptance of ATM, and many system engineers feel that superfast routers that work with variable-length packets could be better than ATM.

ISDN

We saw in earlier chapters that voice is routinely converted to a digital signal for transmission and switching over the telephone network. In this chapter, we see that data already is a digital signal. The synthesis of voice and data on an all-digital communication network thus seems to be the normal progression. This same network could also carry digitized video and digital facsimile signals, since in digital form, all signals are simply bits. Different services require only different bit rates. The overall concept of such an integration of services through an all-digital network is called Integrated Services Digital Network, or ISDN for short.

ISDN ultimately is a worldwide network composed of many smaller private and public networks, all interconnected through common standards and providing a total end-to-end digital capability. This end-to-end capability would extend directly to a standard digital outlet on each customer's premises. At a minimum, three two-way digital circuits would be offered in the following configuration: two B channels each operating at 64 kbps and a single D channel operating at 16 kbps. This minimum configuration is known as the 2B+D basic rate, or the basic rate interface (BRI). The B channels would be used for digital voice and for high-speed data communication, perhaps even simultaneous voice/data communication, and the D channel would be used for signaling and lower speed data communication. Telephone instruments with built-in analog/digital converters would be needed to connect to the basic access interface. The existing copper wire in the local loop could, in most cases, accommodate this basic access to ISDN over a single pair.

A second configuration would consist of 23 B channels, each operating at 64 kbps, and a single D channel operating at 64 kbps. This configuration is known as the 23B+D primary rate, or the primary rate interface (PRI). Twenty-four 64-kbps digital circuits multiplexed together is the standard DS-1 or classic T1 digital signal. Thus, the backbone of ISDN is the standard DS-1 signal at 1.544 Mbps. The European version of ISDN offers 30 B channels and one 64-kbps D channel as their primary rate operating at 2.048 Mbps.

Another form of ISDN that is talked about today is broadband ISDN, in which a digital signal at rates as high as a few hundred megabits per second would be brought to the customer's premises. This broadband capability would be used for such services as high-definition television and two-way picturephone service. More is said about these futuristic services in a later chapter.

The primary driving force for ISDN has been technological in nature. Indeed, the network is mostly digital today, but the local loop is still mostly analog. Why, then, should the digital capability in the network be extended along the local loop to the customer's premises? To answer this question, we need to examine the services that would be made possible with ISDN. The data services most frequently mentioned are electronic mail for the business market and Internet access for the home market. Most large businesses already have their own PBX, which most likely is a digital switching machine. Data communication in many businesses is mostly local and is carried over LANs. Such services as facsimile work over existing analog circuits. Consumers do want high-speed access to the Internet, but modems operating at speeds of 56 kbps on regular phone lines are readily available—ISDN would only approximately double this capacity, and at a much higher cost. We thus see that from a consumer perspective, there is questionable need for ISDN.

Indeed, the network is digital, and in digital form, all signals are the same. Thus, if one defines ISDN as a digital network carrying digital signals representing many different services, then ISDN is already here. But if by ISDN one means an end-to-end digital capability over the local loop, then ISDN is a long way from arriving. In the next century, optical fiber might replace copper wire in the local loop. But this will occur for engineering and economic reasons for the delivery of telephone and data signals and not for the delivery of new services. Clearly, the signals carried over the fiber will be digital, and then ISDN will have arrived in its end-to-end definition. But other ways to deliver digital to the home are available.

ADSL

ISDN is a system to create digital connectivity over copper wire pairs of the local loop to deliver digital data and speech. ISDN requires a single pair of copper wire and is symmetric in the sense that the capacity is identical in each direction. Much data communication, however, is asymmetric, requiring a much higher capacity in the downlink to the home and a much lower rate in the uplink from the home. Access to the Internet is particularly asymmetric, with the user usually sending only a few keystrokes back to the Internet service provider and then receiving tens of thousands of bits in return. The solution to this type of asymmetric communication is an asymmetric digital subscriber line (ADSL).

ADSL can accommodate from 1.5 Mbps to 6 Mbps from the central office to the subscriber and from 16 kbps to 640 kbps from the subscriber back to the central office. Analog voice and digital data are multiplexed together on a single pair of copper wires in the local loop. The voice band is left untouched with ADSL, unlike a modem that modulates data using the voice band. Thus, voice and data can be used simultaneously with ADSL.

ADSL is, in effect, a data modem that is able to operate simultaneously with voice telephony by using a band above voice for data—data above voice. Different approaches are being used to transmit the data signal. In one approach, the data signal is used to modulate the amplitude and phase of a high-frequency carrier, but only when actual data needs to be sent. This is a form of suppressed-carrier quadrature modulation and is called carrierless AMPM. A carrier is sent only when there is data to be sent; otherwise, the data channel is silent, thereby minimizing crosstalk with other lines. In another approach, many carriers simultaneously spread across the entire band are used according to the noise characteristics of the line. This approach is called discrete multitone (DMT).

For ADSL, the loop length must be less than a few miles (usually specified as less than 18,000 feet), and there can be no loading coils on the loop. A splitter in the home separates the baseband analog voice channel from the digital data operating in a higher frequency band. A similar splitter performs the separation at the central office. The analog speech signal goes to the voice switch, and the digital data goes to an ADSL modem dedicated to each subscriber, and then the data signals goes to a multiplexer for access to an appropriate data network such as the Internet. The digital data channel is always active for near instant access.

An entire family of digital subscriber line technology has emerged—referred to as xDSL. High-speed DSL (HDSL) offers T1 capacity symmetrically on two wire pairs. Very high speed DSL (VDSL) offers even greater capacities—as high as 52 Mbps—but works only over short loops less than 1000m, mostly for business customers.

Local area data networks

The widespread use of personal computers as work stations in offices, schools, and other places has created a need to share certain limited, expensive resources, such as laser printers and large disk memories. To accomplish this sharing, a number of computers must all be able to access the same equipment. This can be done through the use of cables and a switch of some kind, but a more flexible way is through the use of a local data network linking together the computers and the shared equipment. Such networks are called local area networks, or LANs. A wide variety of LANs are available, depending on such factors as the network topology, the transmission medium, the multiplexing technique, and the access control protocol. Most LANs use some form of packet switching.

The popular network topologies are the star, the ring, the bus, and the mesh; these topologies are shown in Figure 10.4. In the mesh, each node is connected directly by separate wires to all the other nodes. A switch at each node makes the needed connection for the transmission of data. The mesh topology needs a separate wire for each node and thus is impractical in cases involving a large number of nodes. The star topology has a centralized hub and switch to which all nodes are connected. The star topology is similar to the local loops and central office used for telephone communication. The mesh and star topologies are mostly of academic interest since they are not used for most LANs.

The ring and the bus topologies are similar in that control is decentralized and the same data signal is broadcast to all nodes. The ring is a loop of some appropriate transmission medium that connects from one node to the next. Data signals are passed or circulate around the loop and are usually amplified and repeated at each node. The bus is a single length of some appropriate transmission medium. Each node is linked to the medium at connections called taps. A bus can be visualized as an opened or broken loop.

Many different transmission media are used for LANs. The transmission medium can be copper twisted-wire pair, coaxial cable, or optical fiber. Both coaxial cable and optical fiber require special connectors at the places where the nodes are attached. Twisted-wire pair, on the other hand, is very simple to connect at the nodes and also is the least expensive and easiest to install.

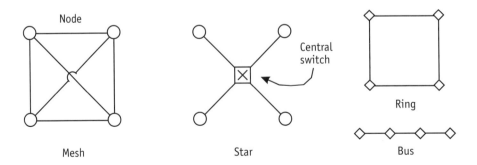

Figure 10.4   Various topologies for networks are the mesh, the star, the bus, and the ring. The last two are used frequently for local area networks.

The data capacities of optical fiber and coaxial cable are higher than twisted-pair wire. Coaxial cable and optical fiber can carry hundreds of megabits per second, while twisted-pair wire can carry only a few megabits per second. Clearly, the choice of transmission medium must consider the data capacity that is needed. A full page of text, for example, is equivalent to about 40,000 bits. If this page is sent to a high-speed laser printer operating at one page per second, then a transmission capacity of less than 50 kbps will suffice, and 100 Mbps would be sure overkill. However, if a number of users are sharing complex graphical information, then transmission capacities on the order of megabits per second indeed make good sense. The capacities of the LAN must be matched to the needs of the users.

With the ring and the bus topologies, all nodes share the same medium, and the signals must somehow be kept separable through an appropriate multiplexing technique. The medium can be shared in time through TDMA or in frequency through FDMA. Time-division multiple access is most frequently used for LANs.

With time-division multiple access, bursts of data can be transmitted over the full bandwidth of the medium in very short intervals of time. A node that has data to be sent, however, needs to know whether the medium is available for use. Various techniques have been devised for controlling this access to the medium. Each node may be polled in turn to determine whether it has data to be transmitted. The problem with this polling method of access control is that some form of central control is needed to perform the polling. Nodes may reserve time slots for the transmission of their data, but here, too, some form of central control of the reservations is needed. The problem with central control is that it adds to the expense and complexity of the LAN, and hence some form of self-control is preferable.

A way for self-controlled access is for each node to check the status of the medium and, if it is not in use, to send the data. Data is sent by modulating a carrier, and hence the status of the medium can be determined simply by determining whether a carrier is or is not present. This type of multiple access by sensing a carrier is called carrier-sense multiple access (CSMA). The problem is that contention can occur if two nodes try simultaneously to send some data, but procedures have been devised to resolve this contention problem.

In one procedure for resolving contention, if the medium is already in use, the node with data to be sent simply waits. The node determines whether the medium is already in use by sensing the presence of the data carrier on the medium. However, two nodes might separately sense no carrier and then simultaneously begin transmission. If this happens, a data collision occurs, and both nodes sense that some other node is transmitting, too. The protocol specifies that both nodes cease transmission and transmit later. An algorithm is used to determine when to attempt retransmission; a random element in the algorithm reduces the probability that both nodes will wait identical times. This access protocol is called Carrier-Sense Multiple Access with Collision Detection (CSMA/CD). The CSMA/CD protocol is used with Xerox's Ethernet™ LAN, which has a bus topology using coaxial cable as the transmission medium.

A second protocol involves a token that is circulated around the network. The "token" is a short data message. If a node has data to transmit, it must possess the token in order to transmit. The node must wait for the token to come into its possession. After transmitting its data, the node releases the token. Alternatively, the node may change the contents of the token to signify that data transmission is occurring and then recirculate the reserved token. Changing the token back to its free state releases the token for use by another node. Token passing, as this protocol is called, is particularly appropriate for the ring topology (called token ring), although it is used with the bus topology, too (called token bus).

With frequency-division multiple access, channels are dynamically assigned and switched to connect together those nodes that are transferring data. This assignment of channels is performed by a master controller. Some nodes may share the same channel, and in this case, some form of contention-resolving procedure is needed. The use of frequency-division multiplexing to share a medium is called broadband transmission.

If a medium is shared in time through time-division multiple access, the data signals are usually transmitted as baseband digital data. The terms "baseband" and "broadband" can be confusing since a medium such as coaxial cable has a broad bandwidth that can be used with either baseband or broadband transmission.

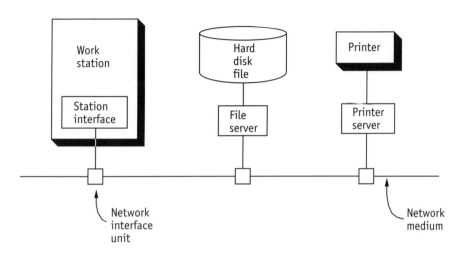

Figure 10.5 Computer workstations, disk files, and printers are connected through two interfaces to a LAN. One interface is the physical connection to the transmission medium of the network. The other interface manages the data traffic and is located at the device being served.

Computers, terminals, printers, and other devices are usually connected to a LAN through two interfaces, as shown in Figure 10.5. One interface is the connection to the network itself through a network interface unit. For a bus network, the network interface could be as simple as an electrical connection to the line by a parallel connection, called a tap. For a loop network, the network interface receives signals sent along the loop, extracts the portion intended for the node it serves, and retransmits the signal along the loop. The network interface unit for a loop network performs as a transmitter/receiver, or transceiver.

The second connection is to the device itself. If the device is a shared resource, such as a printer or a hard-disk file, the interface must manage and serve the access by a number of users. It is thus called either a printer server or a file server, respectively. If the device is a workstation of some kind, such as a personal computer, the interface is called a station interface, or a terminal server, and is usually a printed circuit board, or card, that plugs into the station itself. Servers and station interfaces are microprocessor-based computers that perform communication management functions.

Voice vs. data

The Information Age is upon us, and some futurists predict that the data traffic generated from such services as electronic mail, database access, and electronic funds transfer will soon outstrip voice traffic. Let us examine this seemingly reasonable proposition to see whether it is realistic or preposterous. The examination will estimate the yearly total traffic in bit equivalents for various communication services.

The number of local telephone conversations in the United States in 1995 was 484 billion (*Stat. Abs.*, 1997). Assuming that the length of an average local telephone conversation is three minutes, we have a total of 1,452 billion minutes of use. To this traffic we add 432 billion minutes of interstate traffic and 16 billion minutes of international traffic, for a grand total of 1,900 billion minutes. The speech is encoded into digital form at 64 kbps in each direction, creating a total yearly traffic of 14×10^{18} bits, or 14,000 quadrillion bits (a quadrillion is 10^{15}). We next compute the traffic from various text-based communication services.

In 1996, 290 billion pieces of first-class, second-class, and priority mail were carried in the United States. Assuming that each piece of mail contained the equivalent of one full page of text (4,800 characters at eight bits per character), we have a total yearly traffic, if all this mail were communicated electronically in digital form, of 11×10^{15} bits, or 11 quadrillion bits. There were 1,382 million credit cards of various kinds outstanding in 1996. If each card were used twice a day and each transaction required 200 characters, we have a total yearly traffic of 1.6×10^{15} bits, or roughly 2 quadrillion bits. If we assume that each of the 100 million households in the United States generates or receives two checks a day and that each check if communicated electronically in digital form requires 200 characters, we have a total yearly traffic of 2.3×10^{14} bits, or 0.2 quadrillion bits.

Clearly, the digital sum of these services does not come even close to the traffic generated by voice telephony. Let's assume that the Information Age has truly arrived and that each of the 127 million workers in 1996's labor force work at terminals, entering and retrieving information all day, 365 eight-hour workdays a year. Further assume that they are very productive and generate or use 30 screenfulls of information per hour, where a screen consists of 1,500 characters. This informationcrazed workforce would generate a yearly total of 1.3×10^{17} bits, or 130 quadrillion bits. Voice telephony still wins!

One service that could come close to voice telephony in traffic is the electronic delivery of newspapers. The average length of a newspaper is 100 pages. We will assume that each page contains about 7,000 characters of text (56,000 bits) and about 200 square inches of pictures. Using a resolution of 200 lines per inch and 20:1 compression, this pictorial information is equivalent to 400,000 bits per page. The transmission in electronic form of such a newspaper per day to the 58 million newspaper readers in the United States would generate a total yearly traffic of about 1.0×10^{18} bits, or 1,000 quadrillion bits.

The results of the preceding estimations are shown in Table 10.1

Clearly, the capacity requirements for voice telephony swamp all other communication services. This really is not that surprising since each telephone conversation is a unique message requiring a considerable amount of bits. If speech were converted to text, considerable savings in bits would result, as shown in the following section.

Speech and text

A fairly good rate for reading aloud is about 120 words per minute. If we assume that an average word is six characters long, then the text equivalent of one minute of speech is 720 characters, or 5,760 bits using eight-

Table 10.1
Information Age Traffic

Service	Number of Bits
Telephony	$14,000 \times 10^{15}$
Electronic mail	11×10^{15}
Credit cards	2×10^{15}
Electronic checks	0.2×10^{15}
Information workforce	130×10^{15}
Electronic newspaper	$1,000 \times 10^{15}$

bit ASCII coding. This one minute of speech requires 3,840,000 bits if encoded digitally at 64 kbps. Clearly, text is more efficient than speech in its use of bits by a factor of nearly 700 to 1.

Voice mail is a form of electronic mail in which the actual speech message is stored in digital form for later retrieval by the intended recipient. The digitalization of speech at a rate of 64,000 bps requires large amounts of storage. As shown above, each minute of speech requires 3.8 million bits. However, there is considerable redundancy in the speech wave, and appropriate methods have been developed to reduce the bit rate for storing speech to 9.6 kbps and less. If we assume that the speech can be digitized and stored at a rate of 1.2 kbps, then a one-minute message would require only 72,000 bits. This, however, is still over 10 times the storage needed for a text message in ASCII. Thus, text is more efficient for storing messages, even if compression of the speech signal is used. But we all have telephones—we do not all have text terminals—and nuances of inflection are lost in text messages. Regardless of storage efficiency, voice mail is here to stay.

Text and image

Electronic mail (e-mail) entered from a keyboard and sent as a data signal to the recipient is used frequently in many large businesses. However, the considerable growth of facsimile during the last few years indicates that an easy-to-use, standardized system based on the public switched network is impossible to beat, even though e-mail is more efficient in its use of bits, as we shall soon see.

We assume a full page of text with 60 rows of 80 characters each. These 4,800 characters coded in ASCII require 38,400 bits. This same page of text is scanned by a facsimile unit with a resolution of 200 lines per inch, truly excellent quality. The 8.5 × 11-inch page has about 3.7 million picture elements (pels), and if a single bit is used to encode each pel, then about 3.7 million bits are needed for a facsimile of the page of text. However, there is a great deal of empty space on a page of text, and many patterns appear again and again. Hence, a large amount of compression is feasible for facsimile reproduction. Compression by a factor of 15 is quite possible, thereby reducing the number of bits per page to about 250,000 bits—6.5 times that needed for the ASCII representation. We must remember, though, that facsimile reproduces a handwritten signature, letterhead graphics, and pictorial information—ASCII does not.

A picture is still worth 1,000 words!

In analog forms, a speech signal has a bandwidth of 4 kHz, and a video signal has a bandwidth of 4.5 MHz—a ratio of about 1,000 to 1, thereby proving the old adage that a picture is worth 1,000 words!

There is considerable redundancy in a video signal, both within an individual frame and also between frames. This redundancy can be used to reduce the bit rate required for a video signal in digital form. The digitalization of a video signal can result in a data rate of 90 Mbps, as opposed to a digital speech signal at 64 kbps. The digitized video signal can be compressed to 1.5 Mbps with some degradation of quality, particularly for those portions of the video with considerable motion. The speech signal can also be compressed, perhaps to a data rate as low as 1.2 kbps if some degradation in quality is acceptable. In compressed forms, our ratio of 1,000 to 1 is still preserved.

Data communication—an assessment

Data communication has its forms of quality degradations that must be considered from an engineering perspective. Data routers and switchers introduce transmission delay into the process as a data signal is relayed from a source to its destination. Noise is always present and corrupts signals. The bit error rate (BER) must always be measured and kept as low as possible, although error detection and correction techniques can help assure near perfect transmission.

All the hyperbole about the Internet has warped network architecture toward a packet-switched data model. But data is only the tail of the network—voice is the body. This is not to imply that data communication is not important. A public switched data network is as essential as the public switched telephone network (PSTN). Data communication seems best served by packet switching; voice by some form of virtual circuit switching.

Selected bibliography

Black, U., *ATM: Foundation for Broadband Networks*, Englewood Cliffs, NJ: Prentice Hall, 1995.

Davenport, W. P., *Modern Data Communication*, Rochelle Park, NJ: Hayden, 1971.

DeNoia, L. A., *Data Communication*, Columbus, OH: Merrill Publishing Company, 1987.

Doll, D. R., *Data Communications*, New York, NY: John Wiley & Sons, 1978.

Fisher, W., et al., "Data Communication Using ATM: Architecture, Protocols, and Resource Management," *IEEE Communications Magazine*, Aug. 1994, pp. 24–33.

FitzGerald, J., *Business Data Communications*, New York, NY: John Wiley & Sons, 1984.

Miller, A., "From here to ATM," *IEEE Spectrum*, June 1994, pp. 20–24.

Stallings, W., "A Manager's Guide to Protocols for Local Networking," *Telecommunications*, Sept. 1987, pp. 38–54.

11

Communication Services: Old, New, and Questionable

Introduction

The technology of today's telecommunication network is very impressive and sophisticated, but it is not technology *per se* that consumers purchase. Consumers purchase services and what the technology can do for them. This chapter describes and discusses telecommunication services along with the acceptance and rejection of these services by consumers. Some of these services are old and are well accepted by consumers. Other services, like cellular telephone service, are relatively new but are achieving rapid acceptance in the marketplace. And still other services have had a questionable acceptance and perhaps will continue to have poor prospects for market success.

The telephone has been such a great success that we sometimes forget that telecommunication has not been without its failures, too. These failures are not failures of technology, but rather are failures of markets to appear for some of the new products and services made possible by the technology. In many cases, the market failures occurred because the consumer had little or no use for the products and services. Furthermore, consumer needs and behavior in using novel forms of communication were frequently poorly understood.

This chapter describes and discusses market laggards and outright failures so that we can learn about the nature of human communication, see the future more clearly, and be more realistic in our enthusiasm for technology. Later in this chapter, I describe services that have succeeded. We shall see that the public switched network is a great resource, making possible a host of useful services—services that enable us to summon emergency assistance, to make toll-free calls, to transmit documents, and to reach people wherever they may be.

Lessons from picturephone

It would seem that every industry has its "Edsel"—some technology, product, or service that has failed in the marketplace despite extensive efforts to promote and sell it. The telecommunication industry's Edsel is the picturephone—a telephone that enables you not only to hear but also to see the other party. In return, you are not only heard by the other party but are seen, too.

In the late 1920s, Bell Labs conducted research into a visual communications medium; it would later evolve into television. AT&T demonstrated the results of this research at the Chicago World's Fair in the early 1930s as a two-way, color, video telephone. In the late 1950s, Bell Labs initiated development of a trial videophone. The results of this development effort were demonstrated at the New York World's Fair in 1964 as the picturephone system. AT&T interpreted the consumer response to the demonstration as strong indication of a positive market need, although 700 interviews were conducted only of people who used the demonstration picturephone system, and even then, only one-half of them said it was important to see the other person during a telephone conversation. Thus, in the mid 1960s, AT&T made the decision to introduce picturephone service in 1970. Indeed, after much development work, picturephone service was introduced in 1970 in Pittsburgh, and at that time, AT&T forecast 41,500 picturephone sets in use in 25 cities by 1975. Picturephone service was introduced to Chicago in 1971, and at that time, the Institute for the Future performed a Delphi study that forecast over two million picturephone sets in 1985.

The picturephone system introduced in the early 1970s worked over conventional twisted pairs of copper wire on local loops of less than six miles in length. Three twisted pairs were used: one for two-way audio and two for two-way video. The video bandwidth was 1 MHz—much less than over-the-air television's 6 MHz. The bandwidth reduction was achieved by lower resolution: 250 lines per frame vs. 525 lines for broadcast television. Like broadcast television, 30 full frames were transmitted per second. The picture size was a small 5½ by 5 inches, designed to show only the head and face of the user. Picturephone service cost a flat $160 per month for the first 30 minutes and then $0.25 per additional minute.

The market acceptance of picturephone service was very poor, and in 1972 AT&T cut back on the manufacture of picturephone sets, after having spent an estimated $500 million on the project. Amazingly, AT&T had developed the picturephone system and had introduced it into the marketplace with only the most meager evidence of any real market need—it was mostly hope and belief that guided the picturephone project.

With the poor market acceptance of its picturephone service, AT&T finally initiated extensive consumer research. A study conducted in 1971 showed that strict face-to-face applications of picturephone communication had limited appeal and that "lack of application" was the major reason for not subscribing to the service. However, this study did show that document display had a relatively wide interest. In 1972, AT&T admitted its problems with the picturephone system and announced that it was about to begin a market exploration program to study the market for two-way, switched visual communication. A picturephone system had been installed for use by AT&T management, and the results of a study of these users conducted in 1974 indicated it would make little or no difference to them if picturephone service was discontinued. Two-thirds of picturephone users used it for applications involving graphics, and consumer studies conducted in 1974 indicated that the greatest need was for high-powered graphic communication and that there was "little need for face-to-face communication" over a picturephone system. A study conducted in 1975 of past executive users of picturephone service concluded that it was "a bust." The final conclusion of all these studies was that picturephone service was "a concept looking for a market" and "an idea ahead of its time." With hindsight, we see that the continued expression of need for some form of graphical communication predicted the recent success of facsimile.

As part of its market exploration program, AT&T searched for high-need applications for two-way, switched visual communication. One such application was the picturephone system that was operated at the Bethany-Garfield Hospital in Chicago from March 1973 to April 1976. About 20 picturephone sets formed the system, and usage averaged about 1,800 calls per month. The United States Department of Health, Education, and Welfare funded the costs of the system as part of its research into new telecommunication technologies. Although the system was being used, I was suspicious of why, particularly when I visited the installation and noticed that the sets were often so positioned that the person at the other end received a picture of the user's belt buckle. Hence, my colleague at AT&T, James P. Woods, and I conducted a study to determine why people were using the system.

Interviews were conducted of the 57 hospital people who had access to the system. The results showed that the system was used primarily because of its ability to reach people quickly. Because of its distinctive ring and because it could be used only for internal calls, the picturephone system was, in effect, a "hot line." Another reason for its use was the ability to show forms. The ability to see the other person was a distant third in the order of reasons for use. Only the people who used the picturephone system the least rated the ability to see the other person as its best feature. When the federal funding ceased, the hospital had the system removed.

A picturephone system was installed in Phoenix in the 1970s for use in the criminal justice system so that attorneys could use the picturephone to see and to talk to their incarcerated clients rather than physically going to the jail. The visual demeanor of a client was important to assess for an attorney, and thus the use of the picturephone system seemed to make good sense. However, the same type of service could have been provided by a simple closed-circuit, two-way television system. The flexibility of the switched picturephone system simply was not needed in this application, although it was a good application for two-way video.

The virtually unlimited bandwidth of optical fiber coupled with digital switching and increasingly less costly television cameras and displays create a hope that the technical and cost environment for picturephone service is becoming more positive. This hope assumes that economics was a major cause of the failure of picturephone and that newer technology is the solution. I think not. Picturephone failed for the simple reason that it was not an improvement over a conventional telephone call. The intimacy of a telephone call is shattered by seeing a face on a screen. One must sit still to remain on camera and must be concerned with visual impression. Based on informal study, I am convinced that many people would actually pay extra *not* to have picturephone service if it were standard.

In the early 1990s, small two-way visual telephones that worked over regular telephone circuits were introduced, even by AT&T. These visual telephones used algorithms to compress and reduce the bandwidth needed to transmit the visual image. The resolution of the image was poor, although perhaps just acceptable for seeing a face. These visual telephones failed in the market, just like AT&T's original picturephone, confirming that the consumer did not see value nor have need for such a form of visual communication. But some of my friends and colleagues, who may otherwise appear quite intelligent, continue to believe that the picturephone will ultimately replace the telephone. The latest approach is videophones that work with a personal computer. But why anyone would want to watch me typing away at my computer evades me.

Teleconferencing

The energy crisis occurred at the time of the visual-communication market exploration program, and this led to an investigation of two-way video teleconferencing by AT&T. AT&T's videoconferencing system interconnected such major cities as New York, Washington, San Francisco, Los Angeles, and Chicago. The service was available at public rooms located in AT&T buildings in these cities, and the rooms in any two cities could be connected. In response to customer comments, continual improvements were made in the technology until it finally offered color video at the NTSC broadcast standard, overhead graphics, and full-duplex audio. The rooms were designed to be pleasing and easy to use. Full-time attendants staffed the rooms and were available to assist the users. The rates were set assuming future technological advances to reduce bandwidth and were only about $400 an hour for a coast-to-coast teleconference.

The teleconferencing service was demonstrated to many AT&T customers, and nearly all who saw the system were impressed and were positive toward using it. Some actually came and used it for a meeting or two, but the problem was that few, if any, came back to use it again. I remember a focus-group session in which the participants were all positive toward the service based on a demonstration but had not used the system again. At the end of the session, the moderator stated that AT&T would offer a free call to the participants—not one participant was interested! It was then that I realized we could not even give video teleconferencing away for free! This poor response was replicated with an internal video-conferencing system that was installed in July 1975 between an AT&T building in Morristown, NJ and AT&T's corporate headquarters in Manhattan, NY. A study predicted heavy daily usage of the system, but after six months, actual usage was quite low.

Clearly, the problem with video teleconferencing was not technology or economics. The answer came from research conducted in England and in the United States into the purpose of meetings. The results of this research indicated that teleconferencing was most useful for meetings characterized by the exchange of information between participants who knew each other. Two studies that I conducted indicated that such "target market" meetings are only 4 percent of the total of all types of group meetings. Thus, AT&T's difficulty in promoting video teleconferencing involved the types and purposes of meetings. We also discovered that few people want to use public rooms and facilities.

I personally believe there is a market for teleconferencing of all kinds, from the simplest audio-only systems to elaborate full-motion video systems. One of the most impressive video systems I have ever seen is Bellcore's VideoWindow™ system, shown in Figure 11.1. I advocate a "product line" approach to teleconferencing in which a spectrum of technologies are offered depending upon distance, budget, and types of

Figure 11.1 VideoWindow is a video teleconferencing system developed by Bellcore engineers. Two video images are projected side-by-side using rear-screen projectors and are life size. There is no switching of cameras, and it appears as it the remote participants are sitting there on the opposite side of the room. High-quality, full-duplex audio is also used, facilitating normal conversation. The system is installed between Bellcore facilities in Morristown and Navesink, NJ, a distance of about 70 miles by car. (Photo courtesy of Bellcore.© 1990 Bellcore)

meetings. For many applications, simple, audio-only systems augmented with facsimile are sufficient. Full-motion video is the "Cadillac" of tele-conferencing, and, even with modern bandwidth compression, is still quite costly.

Claims have been made that in 1990 there were over 2,500 public and private video teleconferencing rooms in the United States. Perhaps I should be impressed with this demonstrated "success" of video teleconferencing, but when I think of the vast numbers of public and private telephones, facsimile units, and cellular telephones, I do not get excited about video teleconferencing. The "problem" with teleconferencing of all kinds is that hyperbole and hope far outstrip realistic expectations. It fascinates me that such an overwhelming enthusiasm can still exist toward video teleconferencing, given its history of low market acceptance and use. Yet in 1990, AT&T announced with great enthusiasm the availability of a two-way video teleconferencing service between public rooms in New York and Moscow at the rate of $3,200 per hour. I suspect that it was idle most of the time. At the end of the 1990s, AT&T no longer even remembers it.

Videotex and the Internet

Videotex is an electronic information service for the home that uses the home TV set as a display and connects to the central database over a telephone line. Videotex is not new. The British Post Office spent over $50 million during the early to mid 1970s developing its Prestel videotex service, which was introduced in 1979. The Prestel service was designed to use the home TV set for display of text and graphic information retrieved over telephone lines from the central database. The information was organized in a tree-like fashion to make it easy to find and retrieve a specific page of data. Sports, weather, shopping, finance, and travel information were some of the information services that were available. The British coined the term "viewdata" to apply to their service. The acceptance of the service by British consumers was considerably less than expected.

The United States market for videotex was explored in the early 1980s. AT&T joined with Knight-Ridder Newspapers in a videotex service, called Viewtron service, which was initially offered on a trial basis to participants in Coral Gables, FL. AT&T supplied the terminals to attach to home TV sets, Southern Bell supplied the network, and Knight-Ridder formed a subsidiary—the Viewdata Corporation of America—to supply the database. The service that was used the most was a text-based messaging service, known as bulletin-board, in which the participants could enter and store messages for each other. AT&T and Knight-Ridder became enthused about Viewtron and decided to launch it commercially. However, the consumer acceptance was far less than expected, and the service was withdrawn. At about the same time, AT&T joined with Times-Mirror to develop and offer a similar viewdata service in Orange County, CA. I remember visiting the Times-Mirror videotex office and attempting to use the service to determine whether there was a certain department store in the mall across the street. After spending an hour searching the database, we gave up and used the telephone book. The Times-Mirror videotex service met with low consumer acceptance, and it too was terminated. Similar market failures of videotex occurred around the world in such countries as Canada, Germany, and Sweden.

In 1996, WebTV® service came upon the scene. It offers access to the Internet using an adapter that connects the home TV set to a phone line. In 1997, the company was acquired by Microsoft for nearly $0.5 billion. WebTV service has not yet achieved the great success that it forecast for itself. WebTV looks at lot like videotex, and its lack of great success thus far is hence no surprise. Most people do not want to use the home TV set as a way to view information accessed over the Internet. The WebTV adapter costs a few hundred dollars and there is a monthly charge for access to the service. The TV display does not have the clarity obtained on a personal computer. WebTV service is an oddity that may soon fade away, joining picturephones in the wastebasket of market failures.

In France, a small Minitel terminal, which also can be used as a telephone, enables access to databases containing telephone listings and other information. The French Teletel service is considered a "success," and that is because the French government gives away the Minitel terminals. The Minitel is an integrated telephone and simple black-and-white display terminal. It apparently is easy to use, and the French have created a widespread packet-switched data network to support the system. Millions of Minitel terminals are in use, and the French were actively promoting their system around the world a few years ago. The largest use of the Teletel service is to obtain telephone numbers from the database, and the second largest use is to send text-based messages. In 1988, there was less than one Minitel call per day per subscriber—much less than telephone use in the United States where ten telephone calls per day is the norm. Also, most of the present use of Minitel is by business, which means that home use is probably much less than a call per day. Are Minitel terminals gathering dust in many French homes? Do the French have a unique fondness for the written word and for sending text messages? A study done by the French government accounting office indicated that Teletel was quite costly and had accumulated quite a large loss, despite claims to the contrary by the Teletel people. But the Minitel is a dedicated terminal—not the home TV set—and hence is much closer to a personal computer used to access the Internet.

In the mid 1980s, AT&T, Sears, and CBS formed a joint venture called Trintex to develop a videotex service. After experiencing all the market difficulties encountered in Florida and Orange County, AT&T withdrew from the venture and was replaced by IBM. CBS then withdrew, leaving only IBM and Sears to develop what would become Prodigy^{SM} service. IBM and Sears reportedly spent over $500 million in developing their Prodigy service, which was introduced in 1988. After years of losing money, IBM and Sears finally sold Prodigy in the mid 1990s. Prodigy service was a computerized database containing all sorts of information that can be accessed over the telephone network using a personal computer and a modem.

Computerized database services of all kinds and varieties have abounded in the United States for decades. For decades, both the Source and CompuServe offered computer hobbyists and others a mostly text-based videotex service used for messaging and access to text-based databases. There also are a number of database services used by professionals, such as the LEXIS® service offered by Mead Data Central to legal professionals and the MEDLINE system operated by the National Library of Medicine for physicians and medical researchers. But these databases were awkward and difficult to use. The protocols developed for the World Wide Web changed all that by making it very easy to access a multitude of dispersed databases around the world, jumping from one to another at the click of a mouse button.

The power of the Internet and the World Wide Web is its standardized protocol and the use of a simple language, called hypertext markup language (HTML), for specifying pages of information in a way that is independent of the computer platform. Addressing is stated in a simple way that is even understandable to noncomputer experts. After a few minutes of training, anyone could guess that AT&T's Internet address is http://att.com. The HTML-encoded information is decoded and displayed by software called browsers. We have begun the battle of the browsers with Netscape's Navigator being challenged by Microsoft's Explorer. The Internet and the Web have totally replaced videotex as the way people obtain information from computerized databases and send e-mail.

At one time in the mid 1970s, I too believed in videotex and encouraged AT&T to investigate this market. I suggested that one approach for AT&T to enter this market would be to offer an electronic version of the telephone directory, including its classified section, that could be accessed by a simple terminal. I also believed that people had a need to send text messages to each other—a form of electronic mail. As a strategy to enter the e-mail market, I suggested the development of a text terminal that was compatible with the terminals used by the deaf. But we could not find much interest in such a terminal by the mass market. Now that many homes have personal computers with modems for access to the Internet this situation has changed considerably. But for a fast request for a simple fact, such as a telephone number, it is tiring to turn on my personal computer, load software, and connect my modem. However, if my telephone were also a terminal, like the French Minitel, I might be more willing to use it to access a remote database of telephone numbers and classified information.

It would be nice to know tomorrow's weather or today's traffic conditions without waiting for the radio announcer to finally tell me what I want to know. It would be nice to somehow have instant access to such general, timely information without the bothersome procedure of using a personal computer to access the Internet. Obtaining such information by the home TV set might indeed be the way, although WebTV service seems to make it all too complicated. Although we do not have it in the United States, simple and fast access to such information is possible in Europe and Scandinavia through a system called teletext. It is a success there.

Teletext

Teletext was invented in the 1970s by the British and transmits a few hundred frames of text and graphic information over the air along with the standard television signal. The frames of information are inserted in the vertical retrace interval of the TV signal and are retrieved and displayed on the home TV set using special decoding circuitry in the set. The frames are sent in a round-robin fashion, requiring about 30 seconds to send the complete set of frames. On the average, about 15 seconds is required to retrieve a specific frame.

Teletext is a success in Europe and Scandinavia. In England, nearly every TV set sold there is teletext equipped, and over one-third of TV households have the service. Teletext is free. If it were available in the United States on a widespread basis, one could envision a scheme in which it would be financed by giving local information in support of national commercials. Teletext is easy to use and offers fast access to timely general information of a mass-market appeal, such as weather reports, traffic conditions, TV listings, and news headlines. The TV remote control is used to specify the number of a desired frame of information. Confusion over standards, questions about how to finance its operation, and a fascination with high-definition TV appear to be some of reasons teletext is not available or promoted in the United States. The success of teletext around the world is another example that the United States is no longer the only country at the leading edge of consumer electronics.

Electronic home banking

Electronic teller machines are a great success. Consumers have a need to obtain funds at all sorts of hours, and it is normal to see lines of people waiting at a teller machine late at night to obtain some cash for some purpose or another. The extension of banking to the home through the use of a terminal at the home would seem to make sense, and a number of banks have conducted trials of home banking, frequently in concert with some videotex system. The videotex terminal displays balances and allows the user to move funds from one account to another and to pay bills. Unfortunately, it does not dispense cash!

For me, it is too simple to write a check and insert it in an envelope to pay a bill. I do not have the time to use a terminal to type electronic checks, nor would I trust such a system without keeping my own written records for verification purposes. One needs an easy way to reconcile a credit card statement with purchases made online. My need is an easy way to get cash, and here the home terminal is no help. I guess that my behavior continues to be representative of many other consumers, and most attempts at home banking with a terminal of some kind have met with failure and have been canceled.

However, I would like to be able to determine whether some expected deposit has arrived at the bank, but telephoning the bank to speak to some surly employee is too much of an effort except in the most serious cases. My bank loses deposits and makes other mistakes, and hence I do have a need to verify deposits and balances. One solution is the use of a personal computer and the Internet, and perhaps this is how electronic banking will make it to our homes, but only if it is free. Another simple way for me to check my banking balance is audiotex, and it is here today

Audiotex

With audiotex, a touch-tone phone is used to enter digits, such as the bank account number, and to request various services, such as the account balance. The bank's computer responds by reading back the requested information using recorded or synthetic speech. Audiotex (sometimes spelled audiotext) is easy to use, and no sophisticated training or terminal is required. Audiotex banking has been a great success, one example being Citibank's CitiTouch^{SM} system.

I remember a few years ago, when I returned to campus in September, that the usual long lines of students waiting to register for class were not there. My first thought was that the university had lost all its students and I would soon be without a job! Audiotex was the cause. Students now use audiotex to register for class from home or a dormitory room. At least one airline offers an audiotex system so that people can check on expected arrival and departure times of their flights. The system, like most audiotex systems, is quick and easy to use, and saves labor costs for the supplier. Audiotex is a real winner in my judgment, and I expect to see its use continue to grow. Its success is based on its use of the existing telephone network.

Facsimile

The invention of facsimile, or fax, for short, in the 1840s predates the telephone. In 1977, only about 130,000 facsimile terminals were in use in the United States; today, the number in use is many times that number. Facsimile exploded in use over the past few years. One reason for this success is the ease of use of facsimile; another reason is the familiarity and utility of a paper copy. Facsimile works over the switched public telephone network and thus goes everywhere there is a telephone line. At one time, one vendor's facsimile machine would not work with another vendor's machine. This compatibility problem was solved by the introduction of standards by the International Telegraph and Telephone Consultative Committee (CCITT), and this was a big factor in facsimile's great success.

A facsimile system consists of a scanner that scans and converts a document for transmission and a recorder that reproduces the document at the receiving destination. Usually, a scanner and a recorder are located within a single facsimile machine so that it can be used both for transmitting and receiving documents—such a machine functions as a transceiver. The earliest scanner consisted of a drum around which the document was wrapped. The drum rotated while a screw mechanism moved a photocell along the length of the rotating document. The recorder was similar except that a writing element was used with some form of photosensitive or electrosensitive paper. The process was slow: six minutes was needed to send a single page.

A modern fax machine uses a linear array of photodiodes to scan a whole line at a time. Typically, 1,728 diodes are in the array. An optical system images the line to be scanned onto the diode array. The document is moved along a line at a time until the scanning process has been completed. The output signal from each diode represents the information in a small picture element on the page. An alternative method of scanning is the use of a laser beam: the amount of laser light reflected from the page as it is scanned is sensed. These types of scanners are called flatbed scanners, as opposed to the older drum scanner. If represented as only black or white, a scanned picture element is called a pel; if a grayscale is represented, the scanned picture element is called a pixel.

A number of methods are used for the recorder in a modern fax machine. Plain paper is used with a xerographic process. A laser beam writes the information onto a photoreceptive drum, toner is attracted to the charged portions of the drum, and toner is transferred and fused to the paper by heat. Heat-sensitive paper is used with the thermal recorder. A write head with 1,728 fine wires, corresponding to the diodes in the diode array, heats the paper as it is drawn through the recorder.

The CCITT established three groups of standards for facsimile machines that work over the existing telephone network. A fourth group is intended for use on end-to-end digital networks of the future. The Group 1 standard transmits an $8\frac{1}{2}$ by 11-inch page in six minutes at a resolution of 96 scan lines per inch. A sine-wave carrier is shifted in frequency to indicate white or black. The Group 2 standard transmits a page in three minutes at a resolution of about 100 scan lines per inch. Vestigial amplitude modulation and phase modulation of a 2,100 Hz carrier is used. This more efficient method of modulation allows twice as much information to be transmitted compared to the Group 1 standard.

The Group 3 standard transmits a page in about one minute at a resolution of about 100 lines per inch. This decrease in transmission time is obtained through the use of run-length encoding of each scan line. With run-length encoding, the length of a run of white or of black picture elements on a scan line is encoded. Typical documents were studied to determine average run lengths, and a modified Hoffmann code was then used to encode the run lengths to ensure a minimum average-length code word. Each scan line has a resolution of 1,728 picture elements. Since the encoding scheme examines only a single scan line at a time, it is called one-dimensional coding. A two-dimensional coding scheme that examines two scan lines at a time is available as an option. This scheme is called the relative element address designate (READ) code. It encodes one line relative to the preceding line.

Group 3 machines contain a modem and transmit digital information over the telephone network. Rates as high as 9,600 bits per second can be used on good connections, but the rate decreases automatically on poorer connections to as low as 2,400 bits per second on the poorest connection. A higher resolution of about 200 lines per inch is available for Group 3 machines, but more transmission time is needed.

Facsimile has become quite ubiquitous. Many sandwich shops in large cities have fax machines to receive take-out orders from their business customers. Because of time-zone differences, fax is the only way to communicate with people on the other side of the Atlantic. Hardly a week goes by that someone does not ask for my fax number. Facsimile has become so popular that the receipt of junk faxes has become a problem.

Telewriters

An early predecessor of the facsimile was the telautograph machine, invented by Elisha Gray in the mid 1800s. The telautograph machine enabled the user to transmit handwriting over telecommunication facilities to a distant location—in effect, the telecommunication of writing, or telewriting. Today's version of his invention consists of a conventional pen at the transmitter terminal. As the user writes on a special tablet, the movement of the pen is sensed and is transmitted to the receiver where a second pen transcribes the identical movement on a piece of paper. The system is very simple, yet effective.

Years ago, I installed a telewriter system between my secretary's desk and my office. My secretary picked up my telephone when I was out of the office or when I was speaking on my line. By using the telewriter system, the secretary could write and send me messages that would be recorded instantly on paper on the telewriter receiver in my office. If I was using my telephone and another call came in, I could know immediately who called and the message. This was more efficient that the hand delivery of message slips.

Another application of telewriters is from the sales floor to the stockroom, as shown in Figure 11.2. The salesperson writes the order on a telewriter machine connected to the stockroom. The clerk in the stockroom then retrieves the item and delivers it to the sales floor, perhaps in a small elevator. I saw such a telewriter system in use in a jewelry store on Fifth Avenue in New York City many years ago.

I would imagine that small keyboards and computer terminals will ultimately displace many of the applications for telewriters. But meanwhile, in this day of overhyped technology, it is interesting to see a hundred-year-old technology still in use!

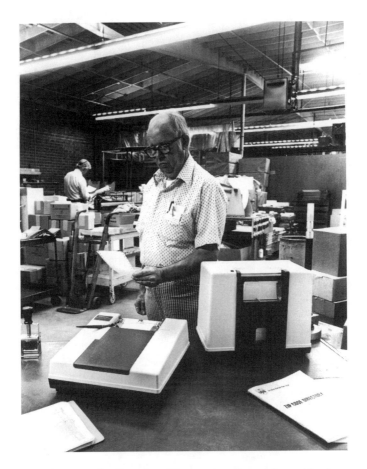

Figure 11.2 Photograph of a telewriting system in use in a shipping department. (Photo courtesy of the Telautograph Corporation.)

Keeping in contact

The airplane version of cellular telephony is GTE's Airfone® service. Nearly two-thirds of the U.S. commercial fleet is equipped with GTE Airfone in-seat telephones. The user removes the handset from the seat-back holder or seat console and selects from a variety of services, such as air-to-ground, ground-to-air, conference calling, data, or facsimile. Billing is accomplished by a credit card, which is swiped at the unit, or through an operator.

The airborne telephone communicates with a network of about 130 ground stations that cover the continental United States, Mexico, and Canada. A total of 310 two-way radio channels are used in the radiofrequency band from 850 to 899 MHz. Each ground station uses 31 of these channels. The ground stations connect to the long-distance network. In effect, the service is much like terrestrial cellular service in terms of the technological principles, except the cells are much larger in size. Worldwide coverage is available on some aircraft through the use of satellite communication.

With today's airborne telecommunication service, passengers can call from the airplane, talk to a colleague in a different seat, check e-mail, or even surf the Internet. GTE announced in 1998 that it was interested in selling its Airfone service, which could mean that the service is not used enough to be profitable.

Many people in the United States subscribe to paging services. We are all familiar with the conventional pager that is worn on one's belt. Newer pagers are so small that they can be worn on the wrist like a watch. Older pagers simply beeped, and the user went to the nearest phone to call the paging office. Newer pagers have small liquid crystal displays of the telephone number of the party to be called back. In some systems, the radio signal sent to the pager is carried on a subcarrier of a local FM radio station; in other systems, the radio signal is sent at a frequency of about 1 GHz. Communication satellites carry paging requests across the continent to metropolitan areas where local paging signals are then sent by terrestrial radio.

800 service

I find it unbelievable, but it is true that about 40 percent of the business calls carried on AT&T's long-distance network are toll-free 800 calls! Though free to the caller, the called party does pay for the call. I use 800 numbers to order goods from catalog stores all over the country. Salespeople on the road use 800 numbers to call their offices to place orders. Today's 800 service is very sophisticated and offers such features as the ability to transfer incoming calls to different offices across the country depending upon the time of day or upon the traffic load.

The 800 service offered by AT&T has a database located centrally in the network. The 800 number is forwarded to that database and is converted to the appropriate routing to reach the called party. The routing information is sent over the common channel signaling network, and the voice connection is finally made.

The 800 service has become so popular that an additional number—888 service—had to be added to indicate a toll-free call.

Emergency services

The 911 service is a real lifesaver. The newest version is called enhanced 911 service, or E911. Calls are automatically routed to the appropriate public safety answering point (PSAP). With the older system, calls might be sent to the wrong PSAP if the boundaries of the telephone exchange did not agree with municipal boundaries. E911 includes a centrally located database of addresses that is automatically accessed when a 911 call occurs. The address information is sent to the PSAP, where it is displayed on a terminal so that the dispatcher knows where to send the emergency response unit. This feature is known as automatic location information (ALI) and is particularly useful if a small child were to call or if the caller did not have time to give the address. With E911 service, the telephone number of the calling party is automatically sent to the PSAP and displayed there, a feature called automatic number identification (ANI). In this way, the dispatcher can always call back the emergency caller if they are disconnected for some reason.

In many states, 911 service is financed by a small surcharge added to the monthly telephone bill. This most certainly is a small price to pay to save a life, but all the cents do add up to millions of dollars. The use of common channel signaling (SS7) in the local network should decrease the cost of providing 911 service in the future.

About one-quarter million people in the United States are protected by the Lifeline® emergency response service provided by Lifeline Systems. This service, and similar systems operated by others, uses a small device carried or worn by the user in the home. If an emergency occurs, the user pushes a button on the device and a radio signal is sent to a receiver connected to the telephone line in the home. The receiver then automatically places a telephone call to a central location and transmits the identity of the emergency caller. The central location then dispatches an ambulance or initiates some other appropriate response.

Television

Although I have my doubts about the market viability of two-way video communication—the picturephone—there is no doubt about the viability of one-way broadcast video communication—television. It was a great commercial success right from its earliest days. I remember when neighbors would all congregate at the family who had the only television in the neighborhood. How will television change in the future? Someday, will television broadcast be digital? Will digital television signals be sent over the air and over CATV systems? This will take a lot of bandwidth, but with optical fiber, the bandwidth is there. Also, image processing by computers could reduce the bandwidth to reasonable amounts.

In 1995, CATV was available to over 95 percent of TV households in the United States, although only about 63 percent actually subscribed. Hence, nearly everyone in the United States has the potential of receiving television over a land-based system. Is television broadcast over the air obsolete? As national policy, would it make sense to require CATV firms to offer basic broadcast television to everyone for free and to charge only for nonbasic services, and what should be defined as "basic" service?

If television is delivered to the home over a broadband medium, such as optical fiber, will the consumer want all the new services that will be possible? One such service is video on demand (VOD)—the ability to request any TV program or video for viewing at any time. We do this now using our VCR and going to the video store to rent a tape. With video on demand, we can stay at home and "dial up" any video. But a tape rents for about $1 per hour; fiber to the home will be expensive, and will we be willing to pay what it might cost to see a video over such a costly medium? Maybe someday the cost will be cheap enough.

Another new TV service is high-definition television. HDTV offers twice the vertical and horizontal resolution of present television, along with a wider screen, but it requires a whole new TV standard along with fairly sophisticated compression to conserve bandwidth. Do you watch

TV because of its image quality or because of program content and quality? Will we be willing to pay a premium of $500 or more for a HDTV set? Will the TV stations be able to afford the costs of replacing all their cameras, recorders, monitors, and other equipment with new HDTV equipment? If the TV set is a conventional screen size and is viewed at its usual distance, will we notice the improvement of HDTV? No, according to research performed thus far. Conventional 525-line TV was designed with the resolution of the human eye in mind and an assumed viewing distance of no closer than four times the screen height. Making the number of lines more will not improve things unless you sit closer to the screen or the screen is much larger. Thus, some people believe that the market for HDTV is a large screen on the wall and the bringing of wide-screen movies into the home—but do you have a whole wall to allocate for use as a TV screen? I, for one, do not.

Perhaps all this excitement about HDTV is fluff. But television at the studio is becoming increasingly digital. Broadcast digital and some form of improved TV or HDTV will soon be available. Perhaps we will have small-screen, portable, personal, HDTV sets that play videotapes along with programs received over the air. But only time will tell if consumers will buy it.

The broadbandwagon

Years ago in the 1970s, it was thought that the coaxial cable medium used by the CATV firms could be made two-way and could then carry a host of two-way services, such as voice telephony and data, in addition to the one-way television signals. Video on demand, in which the user requests some specific video, was also suggested. Some trial systems were installed, but in the end, they all mostly failed. The coaxial cable of CATV is a one-way broadcast medium used primarily for entertainment purposes.

Optical fiber to the home is a broadband medium that today is being promoted for delivering a host of two-way services, such as voice telephony and data, along with one-way television signals. All this sounds quite familiar to the past promotion of coaxial cable. However, coaxial cable is controlled by the CATV firms, and optical fiber thus far is expected to be controlled by the telephone companies. In the past, it was the CATV firms that were looking to services other than the delivery of television signals. Today, it is the telephone companies that are looking to the delivery of services beyond telephone service. A number of policy issues cloud whether telephone companies will be allowed to control the conduit to the home over which television shows will be delivered. The issue of a single broadband medium to the home has pitted the telephone companies against the CATV industry.

From an engineering perspective, the use of a single broadband medium to carry a number of services looks intriguing. However, entertainment video signals are mostly one-way and are broadcast to everyone. Communication signals, such as voice telephony and computer data, are two-way and are sent from one individual to another. The two worlds are quite different. Perhaps the ultimate solution is the use of two broadband media to the home: one carrying one-way television signals and the other carrying a number of two-way communication signals. But meanwhile, now is the time to get onboard the broadbandwagon before you are left behind! We shall see whether it turns out to be another picturephone (a failure) or facsimile (a success).

Selected bibliography

Costigan, D. M., *Electronic Delivery of Documents*, New York: Van Nostrand Reinhold, 1978.

Hunter, R., and A. H. Robinson, "International Digital Facsimile Coding Standards," *Proc. of the IEEE*, Vol. 68, No. 7, pp. 854–867.

Noll, A. M., and J. P. Woods, Jr., "The Use of Picturephone Service in a Hospital," *Telecommunications Policy*, Vol. 3, No. 1, March 1979, pp. 29–36.

Noll, A. M., "Videotex: Anatomy of a Failure," *Information & Management*, Vol. 9, 1985, pp. 99–109.

Noll, A. M., "Teleconferencing Target Market," *Information Management Review*, Vol. 2, No. 2, Fall 1986, pp. 65–73.

Noll, A. M., "The Broadbandwagon," *Telecommunications Policy*, Vol. 13, No. 3, Sept. 1989, pp. 197–201.

Noll, A. M., *Highway of Dreams: A Critical View Along the Information Superhighway*, Mahwah, NJ: Lawrence Erlbaum Associates, 1997.

12

A Short History of Telecommunication in the United States, with an Emphasis on the Bell System

Introduction

For nearly a century, telephone service in the United States was synonymous with the Bell System. Today, the Bell System is no more, although its constituents still exist. A number of firms provide telecommunication service in the United States, although the really big firms have remained mostly the same.

The Bell System had grown during its century of existence until at its zenith in the early 1980s, it had about one million employees, profits over $7 billion and revenues of about $70 billion per year, and total assets of nearly $150 billion. In 1984, this corporate empire was toppled. The story of how that occurred is one of the subjects of this chapter.

Almost as important as the invention of the telephone was the invention of the Bell System to provide telephone service. The structure of the Bell System survived for nearly a century before its fragmentation in the mid 1980s. The forces that led to this fragmentation were technology, competition, and government antitrust action. What is very enlightening is that these three forces were present, in various forms and degrees, from the earliest days of telephony and shaped the Bell System throughout its history. Thus, an understanding of the history of the Bell System and its more recent continuing evolution can help us understand more fully the present, and perhaps even the future, of telecommunications.

It is impossible in one chapter to do justice to all the details and nuances to the 100-year history of the Bell System. The history, then, in this chapter highlights key events that will help us understand the recent and continuing telecommunication policy questions of today.

The invention of the telephone

Alexander Graham Bell is credited with the invention of the telephone. Bell, born in 1847 in Edinburgh, Scotland, came to Canada with his parents in 1870. In 1872, he moved to Boston where he taught teachers of the deaf along with his own deaf pupils. Bell dabbled with early telegraphy and became fascinated with the idea of sending speech directly over telegraph wires. Bell and his colleague, Thomas Watson, worked in Boston to perfect his idea of sending speech electrically, and on February 14, 1876, Bell applied for a patent on this idea. The patent was allowed on March 3, 1876, and issued four days later on March 7. The first working demonstration by Bell and Watson of the telephone did not occur until a few days after that, on March 10, 1876.

Controversy clouds the invention of the telephone. On February 14, 1876, the same day that Bell filed his patent application, Elisha Gray, an electrical researcher and a founder of the Western Electric Manufacturing Company, filed a caveat, or warning, to other inventors concerning a speaking telephone that he was working on. Gray's disclosure included a key concept of a variable-resistance transmitter and was filed only a few hours after Bell's patent application. Bell's application also disclosed the idea of a variable-resistance transmitter, but this idea was written by hand in the margin of the application, almost as if an afterthought. Whether Bell somehow heard of Gray's idea and then added the marginal notation to the application will never be known. In a split decision, the U.S. Supreme Court upheld Bell's claim, but Gray died convinced his ideas were stolen from him by Bell. In addition to his pioneering work in telephony, Gray perfected the telautograph for transmitting handwriting to a distant point, and the Gray National Telautograph Company was chartered in 1888 to pursue his invention.

Bell's teaching of the deaf was closely tied to his financial backers, Gardiner Greene Hubbard and Thomas Sanders. Hubbard had a deaf daughter, who would marry Bell on July 11, 1877. Sanders had a deaf son who was Bell's pupil. On July 9, 1877, the Bell Telephone Company of Massachusetts was formed by Hubbard, Sanders, Bell, and Watson to promote the telephone commercially. Bell's telephone was an instant success and attracted a considerable amount of favorable attention. In the fall of 1877, the early Bell Company had 600 subscribers to their private-line service. By the end of the year, they had more than 3,000 subscribers.

Competition almost immediately characterized the early days of telephony. In December 1877, the Western Union Company created the American Speaking Telephone Company to sell telephone service, utilizing technology invented by Elisha Gray and Thomas Alva Edison. Western Union claimed that their telephone service offered better quality than Bell's.

Early telephone service was offered only as private-line service. That changed when, in early 1878, exchange telephone service was offered for the first time, and a telephone could reach others within the same exchange. However, the Bell telephones were not interconnected with the Western Union telephones, and hence two exclusive telephone systems existed in each area—not a particularly efficient way of providing telephone service.

The formation of the Bell System

Bell needed more funding for additional expansion, and in June 1878, the Bell Telephone Company of Massachusetts was reorganized to create the

Bell Telephone Company. The Bell Telephone Company granted licenses to other companies to install and sell telephone service in all places other than New England, which was the province of Bell's New England Telephone Company, formed earlier that year. As of July 1, 1878, there were 10,755 Bell telephones in service. The early years saw many such reorganizations for the purpose of infusing new capital into the rapidly expanding telephone business. A person later to play an essential role in the formation of the Bell System, Theodore N. Vail was hired at 33 years old as the general manager of the newly formed Bell Telephone Company.

Theodore Vail began his career as a mail clerk in Omaha for the Union Pacific Railroad and moved to Washington, DC to work for the Post Office Department. He became general superintendent of the Railway Mail Service. Hubbard was a member of the Congressional Postal Committee and met Vail. The Vail family had other involvement in telecommunication. Theodore Vail was the second cousin of Alfred Vail, who together with Samuel F. B. Morse, gave the first public demonstration of an electromagnetic telegraph. It is thought that Alfred was responsible for the invention of the dot-dash system of code traditionally credited to Morse.

The competitive battle between the Bell Company and Western Union soon reached a peak, and in 1878, Bell sued Western Union for patent infringement. At about the same time, Bell received rights to the transmitter invented by Francis Blake, Jr., which was better than Edison's. On February 17, 1879, the National Bell Telephone Company was formed from the other two Bell companies. The patent suit against Western Union was settled on November 10, 1879. Western Union gave up all its patents along with 56,000 telephones in return for 20percent of Bell rentals for the 17-year life of Bell's patents. Although this might appear as a costly settlement for Bell, all competition was eliminated, thereby giving Bell a monopoly on telephone service. In late 1879, telephone numbers were used for the first time.

With Western Union eliminated as a competitive threat, attention was turned to structuring the telephone industry. On March 19, 1880, the American Bell Telephone Company was incorporated, and Vail became its chief operating officer. At this time, Vail was beginning to form the framework of the Bell System as a three-legged monopoly consisting of financial control of the operating companies providing local telephone service; the long-lines department of American Bell, which provided long-distance telephone service; and ownership of a manufacturer providing equipment solely for the Bell System. This framework was a vertically integrated monopoly with strong, central supervision by the parent company, American Bell. With this framework as his objective, Vail then initiated steps to bring it together.

The manufacturer that Vail intended to acquire was the Western Electric Company. Western Electric was formed by Elisha Gray and Enos N. Barton, and it manufactured telegraph and telephone equipment for Western Union. Vail began buying its stock and took controlling interest in November 1881. On February 6, 1882, Western Electric and American Bell signed an agreement in which Bell gave patent rights to Western and Western agreed to supply products to Bell exclusively. American Bell determined the specifications for the telephone equipment that the operating companies were to use, and, of course, Western Electric manufactured its equipment to these specifications. In effect, the operating companies were captive to purchasing all their equipment from Western Electric. Later, in 1908, as a response to antitrust agitation, Western Electric agreed to sell its equipment outside the Bell System also.

The first commercially successful long-distance telephone connection took place on January 12, 1881 between Boston and Providence, Rhode Island, a distance of 45 miles. The provision of long-distance service was to be the glue to hold together the Bell System as a monopoly. Vail realized that his monopoly framework would be difficult to dictate after the basic Bell patents expired. In 1884, he developed the argument that control of the long-distance toll network could be the key to retaining the monopoly after the patents expired. On February 28, 1885, the American Telephone and Telegraph Company (AT&T) was chartered with Vail as president. AT&T had control over the long-distance lines, and Vail's structuring of the Bell System was now complete. The reason for the word "telegraph" in AT&T would become obvious about 20 years later as AT&T extended its monopoly to include telegraph service, too.

In 1887, with his structuring of the Bell System complete, Vail left AT&T. A reason given was his disagreement over the strategy to be followed after expiration of the patents and also over too much emphasis on dividends rather than quality of service. Vail spent the next 20 years in semiretirement sailing, investing, and visiting England. The continued growth of telephone service was phenomenal. In early 1881, there were 133,000 Bell telephones; in 1892, there were 240,000; and in 1899, there were more than 800,000 Bell telephones in service.

Vail's dream of a total monopoly notwithstanding, the independent, non-Bell telephone companies had not withered away. In 1893, the basic patents expired, and many non-Bell independents entered the telephone business. In 1900, there were roughly 856,000 Bell telephones compared to 600,000 independent telephones. In 1903, the scale tipped, and there were 2,000,000 independent telephones as opposed to 1,278,000 Bell telephones. This had happened because the independents offered better service and lower rates. The concerns that led Vail to leave AT&T were indeed destroying the Bell System. In response to this competition, AT&T dropped rates and also refused to interconnect the independents to the Bell long-distance network. These policies had an effect, and in 1907, there were 3,132,000 Bell telephones to 2,987,000 independent telephones. However, this small victory was obtained at the cost of a poor public image, and Bell still provided poorer service.

The great monopolist, J. P. Morgan, had gained financial control of the Bell System, and in 1902, the Morgan interests placed Vail on AT&T's board of directors. On May 1, 1907, Vail again became president of AT&T. Morgan and Vail clearly knew each other well and shared the goal of making AT&T the sole supplier of all telecommunication—telephone and telegraph service—in the United States. Together, they bought up the independents. Morgan would cut financial credit to the independents, and Vail would then buy them with a cut-rate deal. The many independents thus acquired were combined with the Bell companies. In 1911, the telephone companies were consolidated and organized essentially as they would remain until divestiture in 1984.

Vail believed that the provision of telephone service was a natural monopoly, but he also recognized that the times were changing and that unbridled monopoly was becoming less acceptable to the public and the government. Therefore, he realized that some form of regulation of the telephone monopoly would be necessary to assure the public and the government that there would be a check on the potential abuse of monopolistic power. State commissions were formed—and welcomed by Vail—to regulate telephone service. By 1915, many states had them, and by 1922, 40 of the 48 had them. These commissions were inadequately staffed and were really little more than rubber stamps of what the telephone companies wanted to do. From all external appearances, however, there was regulation, and the public thought itself protected from the abuses of monopoly.

AT&T extends its power

The Morgan-Vail vision of a monopoly of all telecommunication slowly became a reality. In 1909, AT&T acquired 30 percent of Western Union stock, and in 1910 Vail became president of Western Union. Bell's old foe was now part of the Bell empire, and it seemed that Morgan and Vail were about to achieve their goal of a total monopoly, but forces were in motion that would stifle that goal. In 1910, the Interstate Commerce Commission (ICC) was given jurisdiction over the telephone industry, and it began an investigation to determine whether AT&T was attempting to monopolize telecommunication. Furthermore, in 1912, the remaining independent telephone companies protested to the U.S. Justice Department that AT&T was violating the antitrust laws. As a result of these actions, in January 1913, Attorney General George W. Wickersham advised AT&T that certain planned acquisitions were in violation of the Sherman Antitrust Act.

Fate intervened in the plan to monopolize all telecommunication. Morgan died in March 1913, and Vail had lost his ally. The AT&T response to the Justice Department came on December 19, 1913 in the form of a letter from AT&T vice president Nathan C. Kingsbury to then Attorney General James McReynolds. In this letter (known as the Kingsbury Commitment), AT&T agreed to dispose of its Western Union stock, to purchase no more independent telephone companies (except with ICC approval), and to interconnect independent telephone companies to the AT&T long-distance network. The Morgan-Vail dream of a total monopoly of all telecommunication was dead. Furthermore, the Kingsbury Commitment set a precedent of compromise by AT&T in the face of government intervention, rather than allowing the government to dictate terms.

In 1918, Bell had 10 million telephones. A year earlier, the United States had entered the World War. Since telecommunication was deemed essential to the war effort, in July 1918, President Wilson acted on the authority previously given him by Congress to assume control of telecommunication companies by placing them under the direction of the Post Office Department. For the following year, the federal government controlled the Bell System. During that time, rates rose sharply and a service connection charge was instituted. The World War ended in 1918, and on August 1, 1919, the telephone companies were returned to private ownership, thus ending the government's control of the Bell System, much to everyone's relief. In June 1919, Vail resigned as president and became chairman of the board of AT&T. He died about one year later. To this day, he is revered within AT&T and the Bell telephone companies for the leadership and sound management he gave to the Bell System.

The Kingsbury Commitment was, in effect, given legislative approval by the passage of the Graham-Willis Act in 1921 by Congress. The act formally exempted telephony from the Sherman Antitrust Act and allowed AT&T to consolidate independent telephone companies that competed with Bell in the same geographic area.

AT&T's attempts to monopolize telecommunication extended to other industries, too. In 1922, AT&T owned its own radio broadcasting station, based on its radio patents. Through the use of the telephone network, a national network of radio stations was created. AT&T impeded competition by refusing to allow rival broadcasters to use the telephone network to create their own competing radio networks. Again, the government was forced to intervene, and in 1925 the government threatened antitrust action against Western Electric. Walter S. Gifford, the then-president of AT&T, believed that AT&T should emphasize the provision of telephone service rather than these other business activities. Accordingly, in July 1926, AT&T sold its entire radio network to the Radio Corporation of America (RCA) for $1 million, divested its holdings of RCA stock, and agreed to supply wire services to RCA. In other actions, AT&T sold its Western Electric manufacturing plants outside the United States to the recently formed International Telephone and Telegraph Company (ITT). In 1928, AT&T sold its Graybar Electric division, which was responsible for marketing Western Electric products to customers outside the Bell System.

Interestingly, as AT&T was abandoning the radio business, it was at the same time moving into a monopoly position in yet another entertainment business: motion pictures. In 1925, Bell Labs engineers learned how to synchronize sound with moving pictures, and in 1927, Western Electric sound equipment was used to make the *Jazz Singer*. In 1930, 90 percent of all talking motion pictures used Western Electric equipment.

Meanwhile, the telephone business continued to grow and grow. At the end of 1930, there were 15,193,000 Bell telephones in service. In October 1930, AT&T purchased the Teletype Corporation.

The Bell System

The organizational structure of the Bell System was determined nearly three-quarters of a century ago and persisted until the restructuring that occurred in early 1984.

The Bell System consisted of four major units operating as a single, vertically integrated monopoly, all under the ownership and control of the American Telephone and Telegraph Company (AT&T). This structure is shown in Figure 12.1. The most visible unit to the consumer was the local telephone company that was responsible for operating and providing telephone service, including the leasing of telephone instruments, the wiring of premises, and the provision of dial tone and billing. The total package was bundled together as an end-to-end service with ownership of all facilities by the telephone company, as shown in Figure 12.2.

Twenty-two operating companies were in the Bell System and provided local telephone service. Many of them were corporate giants by themselves, and were household names throughout the United States. New Jersey Bell, Southern Bell, Pacific Bell, Michigan Bell, New England Telephone, and Southwestern Bell were a few of the Bell operating companies—the so-called BOCs—in the old Bell System.

Long-distance service between states is called interstate toll service. It was provided solely and transparently by the Long Lines Division of AT&T. A telephone subscriber making a long-distance call simply dialed the number and had no idea that Long Lines was involved in the call.

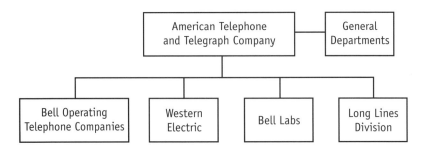

Figure 12.1 The old Bell System consisted of four major operating units, all under the control and direction of AT&T.

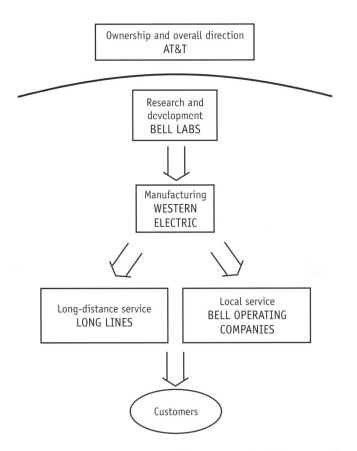

Figure 12.2 The Bell System at its zenith was a vertically integrated monopoly covering all aspects of providing telephone service, including R&D, manufacturing, and the provision of local and long-distance telephone service.

Billing was handled by the local telephone company, and Long Lines was the only long-distance company in town—competition had yet to come to long-distance service. The local phone company gave all the money collected for long-distance calls to AT&T. To compensate the local companies for use of their local facilities in completing long-distance calls, AT&T returned more than half of the monies collected back to the local telephone companies—a process called separations. The local companies relied on this source of revenue to keep the rates for local service low and affordable.

The transmission and switching equipment needed by the Bell operating companies and by Long Lines was manufactured for them on a sole-source basis by the Western Electric Company. The trusted telephone instrument, which steadfastly survived falls on a hard floor, angry slamming of the handset, and spilled cups of coffee, was also manufactured by Western Electric and sold to the telephone companies for installation in the homes and offices of their customers. Western also made such items as the wire used in the provision of telephone service, the various electric components used in transmission and switching systems, and the batteries used to power the telephone system. Western made virtually everything, and the few items it did not wish to make were manufactured to Western specifications by a small cadre of chosen subcontractors.

All the items manufactured by Western Electric were developed and designed by engineers at Bell Telephone Laboratories. Much of the basic technology behind these designs flowed from research performed at Bell Labs. The quality of this research led to the high regard with which Bell Labs was held in the scientific and technological communities.

The stock of the phone companies and Western Electric was owned and held by AT&T. Bell Labs was owned jointly by Western Electric and AT&T, supposedly to acknowledge the relevance of much of the work at Bell Labs to Western's manufacturing mission. AT&T was thus a holding company, but its role extended far beyond being only a silent owner. AT&T was the "conductor," ensuring that the various "players" in the Bell "orchestra" all played together harmoniously. This leadership and direction role was performed by AT&T's General Departments and by AT&T senior management.

The AT&T General Departments addressed problems, issues, and questions common to all the telephone companies in providing telephone service. The work performed by the General Departments was paid for by the Bell operating companies through a levy on their earnings, called the license contract. Ultimately, like all costs of doing business, the license contract fees were paid indirectly by the telephone subscribers. The General Departments also gave direction to Western Electric, and thus telephone subscribers were in effect helping to subsidize Western Electric. This became an issue as Western's competitors entered manufacturing and equipment provision.

The AT&T empire extended beyond telephone service. Nassau Smelting and Refining reclaimed precious and useful metals from scrap provided by the telephone companies from obsolete equipment and facilities. Sandia Laboratories, which was formed after World War II, developed nuclear weapons under contract to the federal government.

The final judgment

The ICC was never that effective in regulating telecommunication, and in 1934 its telecommunication regulatory role was replaced by the creation of the Federal Communications Commission (FCC). The FCC almost immediately initiated an investigation of AT&T, but World War II intervened, with the Bell System playing an important part in the war effort in such areas as radar, sonar, and telecommunication. Thus, no definitive action occurred in terms of the FCC investigation until January 14, 1949. On that date, charges were filed in the U.S. District Court in Newark, New Jersey against AT&T for violations of the Sherman Antitrust Act. The case noted the "absence of effective competition" in telecommunication and asked for Western Electric to be divested from AT&T, for the operating companies to purchase their equipment on a competitive basis, and for patents to be licensed to all for reasonable royalties.

The progress of the antitrust case was extremely slow, and the government did not request documents from AT&T until August 1951. Dwight David Eisenhower, a Republican, was elected President in 1952, and the Korean War intervened. Also, AT&T had involved itself in a very strange undertaking for a telecommunication firm, namely, the design and production of nuclear weapons—a task that continued until 1993. Sandia Corporation was created in September 1949, at the request of the federal government, to take over the design and production of nuclear weapons previously done under the management of the University of California. The slowness of the government in pursuing the antitrust case might well have been an inducement to get AT&T to agree to managing the development of nuclear weapons. Sandia Corporation was a subsidiary of Western Electric until 1993 when the Martin Marietta Corporation (now the Lockheed Martin Company) acquired it, thereby finally terminating AT&T's nuclear history.

On January 12, 1956, the antitrust case was settled by a Consent Decree. The Final Judgment, written by AT&T lawyers, stated that Western Electric would remain with AT&T, and the Bell System would remain intact. AT&T abandoned the audio business by having Western Electric sell its audio manufacturer, Westrex. Furthermore, Western Electric agreed to license all its patents and to manufacture equipment only for the Bell System and not to seek other markets. AT&T agreed that the Bell System would engage only in common-carrier communication and other incidental businesses. Thus, through compromise, the Bell System remained intact, and the government's attempt to pry Western Electric from the Bell System failed again.

Competition comes to telecommunication

AT&T clearly had won again in its battle with the Sherman Antitrust Act, but a new threat in the form of competitive opportunity created by new technology was about to knock at AT&T's door. In 1955, the FCC allowed Hush-A-Phone devices to be attached physically to telephone handsets. The Hush-A-Phone device fitted over the telephone transmitter and prevented one's speech from being picked up. It was simply an acoustic filter. The operating telephone companies refused to allow such devices to be attached to telephones, claiming that the devices were "foreign attachments" not allowed by telephone tariffs. The FCC did not agree and allowed the attachments, but in 1956, the U.S. Court of Appeals set aside the FCC decision. This was not the end of the attachment question, however.

The next threat to the Bell monopoly, which proved to be relevant to its ultimate fragmentation, came in the 1960s from a small mobile radio communication company in Texas operated by the Carter Electronics Corporation. Carter's customers wanted to use their two-way radio equipment to connect with the telephone network to complete telephone conversations. Carter made an acoustic connection between a telephone handset and the radio equipment by using a device called a Carterphone. The phone company objected, citing the prohibition against any "foreign attachment" to its lines or equipment. At this time, even attaching a piece of cellophane tape to a telephone could bring the wrath of the phone company in response to this foreign attachment to its property. Carter was doing far more than that.

In 1968, the Federal Communications Commission decided that telephone tariffs barring connection of customer-owned equipment were unlawful. This decision initiated a series of telephone company actions aimed at impeding and delaying direct electrical connection to their lines. For a while, any direct electrical connection had to be made through the use of a protective-connecting arrangement (PCA) provided by the telephone company. In 1974, customer owned equipment was allowed to be connected directly only if it contained special protective circuitry designed by Bell Labs. The rationale for these actions by the telephone companies were the claims that any direct connection might result in harm to the network and that the telephone companies had a mission to protect the integrity of the network. I doubt if there was any real documented case of any such harm having ever occurred as a result of direct connection of any customer-owned equipment.

In July 1977, the FCC attempted to rectify the situation. Customers were buying telephone answering machines and telephones and were connecting them directly to telephone company lines without any protective circuits. The FCC created a registration program in which the manufacturers of the equipment were required to show that their equipment met certain minimal standards. A certification number was then issued by the FCC.

The problem then became that the telephone company could no longer tell whether telephones connected to the line belonged to the phone company or to the customer. The telephone company then required customers to call the company whenever the customer connected telephone equipment to the line and to report the FCC certification number of the equipment. The certification number contained information about the amount of current drawn by the equipment, and the phone company monitored the total current to determine what was connected to the line. The phone company even attempted for a while to collect a monthly line extension charge for the customer-owned equipment connected to the line.

Another signal that the days of the Bell monopoly were about to end came in 1962 with the creation by the government of the Comsat Corporation to handle telecommunication by satellite. AT&T had claimed that satellite communication was simply another from of microwave radio communication and should therefore be controlled by AT&T as just another part of its long-distance network. However, aerospace people saw satellites as part of the space program and an area of competition by private industry. At one time, AT&T actually asked the National Aeronautics and Space Administration (NASA) for exclusive rights to satellite communication, but NASA refused the request. Comsat was created by the government to be half owned by the communication industry and half by private investors. AT&T was allowed a maximum ownership of 27.5 percent. Comsat became very profitable. In 1972, the FCC pressured AT&T into selling its Comsat stock.

In addition to the attack on the telephone terminal market, AT&T's monopoly on long-distance service also came under attack in the 1960s. Microwave Communications Incorporated, now MCI, installed a private-line microwave system between Chicago and St. Louis. The system did not remain only private line, and it became interconnected with Bell local lines at each end and offered direct competition to AT&T's intercity toll network. After many court and FCC actions, the FCC finally allowed such specialized common carriers in May 1970. Clearly, as AT&T correctly claimed, there was little if anything "specialized" about the long-distance services that these competitors offered. But is was equally clear to nearly everyone, if not to AT&T, that the government's policy was to stimulate competition in the telecommunication industry.

Because private-line service is a specialized offering, as opposed to the offering of conventional, public long-distance service, the companies offering such private-line services were called specialized common carriers (SCC). Their existence was strongly opposed by AT&T, and a lengthy series of FCC and court actions ensued. The whole topic really exploded when MCI attempted to purchase connections to local telephone company lines to obtain access to the public switched network. In this way, subscribers to the MCI service could make a local call in Chicago to an MCI number, dial the desired number in St. Louis, connect via MCI facilities, and then obtain final connection over telephone company lines in St. Louis. MCI called this "new" service Execunet. AT&T's public long-distance service was clearly under attack.

In reality, the only reasons MCI was able to compete with AT&T were (1) MCI did not subsidize local service by paying large fees to the local telephone companies for access to the local network and (2) MCI did not have to provide service to less lucrative cities and areas. AT&T's arguments against MCI were legitimate, but AT&T was viewed as a "nasty monopoly," simply protecting its position. AT&T turned to the political arena to obtain relief.

The Consumer Communications Reform Acts (CCRA) was written by AT&T and introduced into Congress in 1976. A massive lobbying effort was then initiated by AT&T, urging employees and shareowners to write to their elected representatives to vote for the proposed legislation. The CCRA would have wiped out all competition and would virtually have returned to the pre-Carterphone days of the monopolistic past. Congress knows all about lobbying and recognized the CCRA for what it was. However, the result of the introduction of the CCRA was that the Congress took a more active interest in telecommunication policy and legislation favoring competition started to appear. AT&T then found itself actively lobbying to extricate Congress from telecommunication policy. AT&T never should have introduced the CCRA; it was a serious blunder and indicated how panicked AT&T's management at that time had become.

In 1980, the FCC issued the results of its second investigation into computers and telecommunication. In Computer Inquiry II, the FCC stated that all telephone terminal equipment had to be detariffed by early 1982. Furthermore, AT&T was required to separate the provision of regulated services from that of deregulated services. Regulated services were called basic, and deregulated services were called enhanced. "Plain old telephone service," or POTS, is a basic service. Enhanced services pertains to applications and equipment that entail a resale value component, and this is also denoted as value-added service. Confusion between the two types of services still abounds.

The FCC clearly was attempting to promote competition in telecommunication and to force AT&T to reorganize, possibly divesting its manufacturing business. The FCC really did not have the authority to force AT&T to reorganize or to divest any business units, but as AT&T was continuing to lose most court cases, the FCC grew more bold in its dealings with AT&T.

Divestiture—the Bell breakup of 1984

AT&T's attempts to stifle competition during these years had not gone unnoticed by the government. Although apparently a surprise to AT&T, on November 20, 1974, the Justice Department filed its antitrust case against AT&T in the U.S. District Court in Washington, DC. The case asked for divestiture of Western Electric and some form of separation between the operating companies and Long Lines. AT&T, Western Electric, and Bell Labs were named as the defendants, and the operating companies were named as coconspirators.

The case dragged on over the years, with AT&T facing great uncertainty about its future at the same time that competition was nibbling away at its basic telecommunication business. AT&T's management had come to believe that AT&T's future survival depended on it having the freedom to enter new high-technology businesses, such as computers, rather than restricting itself to only common-carrier communications, as agreed in the 1956 Final Judgment. AT&T's historic reaction to fight competitors in the courts and regulatory bodies was consistently failing—it appeared that everyone was now against AT&T.

AT&T entered the early 1980s as a besieged company with many battle wounds and a very uncertain future. On one hand, competitors were allowed to do whatever they wished, but AT&T was still subject to regulation. And to make matters worse, the Justice Department antitrust case was still continuing in Judge Harold Greene's court, with more uncertainty about its final outcome. The FCC was insisting that AT&T entries into unregulated business should be kept fully separated from its regulated business. Hence, rather than wait for the antitrust case to come to a final decision, AT&T settled the case "out of court." The historic precedent of compromise was followed once again.

AT&T's management then saw a vision. That vision was an AT&T free to enter new businesses and markets based upon the scientific and technological inventiveness of Bell Labs. That vision saw the coming of the Information Age and an AT&T free to manufacture and provide computers and other information-age products and services. That vision seemed to exclude plain old telephone service as being mundane and the way of the past. Furthermore, the local telephone companies were being viewed as losers with high costs and declining business. Near panic over what might happen in Judge Greene's court, uncertainty about the future, and a new vision of the future motivated AT&T's management to seek a settlement to the antitrust case.

The Justice Department and AT&T reached a compromise agreement in private and then filed the settlement on January 8, 1982 at the District Court in Newark, even though the then-current case was being heard in Washington. The Newark venue was chosen because the settlement was written as a modification of the 1956 Consent Decree's Final Judgment—thus the term Modified Final Judgment (MFJ). The Final Judgment of 1956 was vacated in its entirety and replaced by the Modified Final Judgment. The clear tone of the MFJ was that the operating companies were responsible for anticompetitive practices and were to be punished by being divested from AT&T ownership and by being restricted to providing only local exchange service. They were not to be involved in manufacturing or in providing long-distance service—areas that AT&T clearly wished to keep for itself without creating any new competitors. Western Electric remained with AT&T, and they both were rewarded by being freed from all the constraints of the 1956 settlement.

Since the original 1956 case, the Congress had passed the Tunney Act, which stipulated that the fairness of antitrust settlements had to be determined by public hearings. AT&T did not want this type of scrutiny nor continued uncertainty and hence attempted to avoid the Tunney Act by making the settlement a modification of the 1956 case. The maneuver failed, and Judge Harold H. Greene angrily had the case returned to his court in Washington.

On August 24, 1982, Judge Greene approved and entered the Modified Final Judgment. He suggested some changes that were agreed to by AT&T and the Justice Department—although one gets the distinct impression that Justice would have agreed to anything since they too were tired of the case and had actually attempted to drop it during its course. The operating companies were allowed to provide (but not manufacture) customer-premises equipment and to publish yellow-page directories. At the instigation of the newspaper industry, which was convinced that AT&T was about to monopolize electronic publishing, AT&T was forbidden from engaging in electronic publishing over its own transmission facilities for the next seven years, after which time the ban would be reexamined.

Under the stipulations of the MFJ, the United States was divided into 160 local access and transport areas (LATAs) that were to be the domain of the Bell operating companies (BOCs). The BOCs were forbidden from providing inter-LATA service and were allowed only to provide exchange service with a LATA. The provision of wireless cellular telephone service was declared an intra-LATA service to be provided by the BOCs.

A controversy developed over the use of the Bell name. AT&T claimed ownership, but it was later decided that the Bell name should be reserved for use by the operating companies. Bell Labs was the only AT&T entity that was allowed to continue to use the Bell name, but had to indicate the AT&T affiliation by calling itself AT&T Bell Labs.

Before the actual divestiture occurred, AT&T reorganized the operating companies into seven regional holding companies (RHCs) and then simply divested the stock of the RHCs. Prior to divestiture, AT&T also created a central organization to perform common engineering and applied research for the regional companies—this organization became Bell Communications Research, or Bellcore.

As a result of divestiture, the BOCs became all independent of AT&T, as shown in Figure 12.3. The stock of the operating companies formerly owned by AT&T was divested and given to seven regional holding companies (RHCs), which then owned the BOCs. The seven regional holding companies and the BOCs that comprised them were as follows:

- Ameritech
  - Illinois Bell
  - Indiana Bell
  - Michigan Bell
  - Ohio Bell
  - Wisconsin Bell
- Bell Atlantic
  - Bell of Pennsylvania
  - Chesapeake and Potomac Companies
  - Diamond State
  - New Jersey Bell
- BellSouth
  - South Central Bell
  - Southern Bell
- NYNEX
  - New England Telephone
  - New York Telephone
- Pacific Telesis
  - Pacific Bell
  - Nevada Bell
- Southwestern Bell
  - Southwestern Bell
- US WEST
  - Mountain Bell
  - Northwestern Bell
  - Pacific Northwest Bell

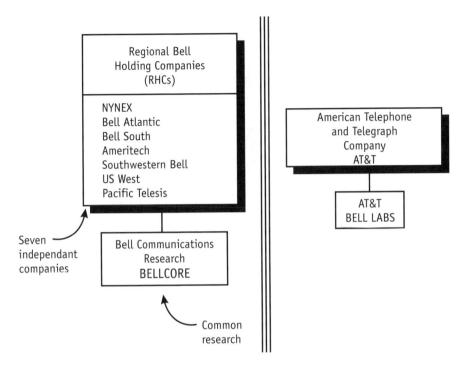

Figure 12.3   As a result of the divestiture that occurred in 1984, AT&T and the Bell operating companies are totally independent of each other. The stock of the BOCs is owned by seven regional holding companies. Bellcore is owned by and performs common research and other activities for the seven RHCs.

Thus, AT&T's monopoly of telecommunication in the United States and the Bell System finally came to an end. The actual divestiture of the operating companies occurred on January 1, 1984. The operating companies all became independent of AT&T, as some of them were before being acquired by Vail nearly a century earlier. But the local operating companies were now in the control of the regional holding companies (RHCs)—the so-called Baby Bells. Some of the regional companies retained the Bell name, and others did not. Southwestern Bell Corporation, the RHC for Southwestern Bell, changed its name to SBC Communications Inc. in 1995.

Not only has the Bell System changed, but the Bell logo has evolved considerably over the years, as shown in Figure 12.4, finally culminating

Figure 12.4 The Bell logo has evolved over time, but prior to divestiture, the "bell" was always an essential feature. After the Bell breakup of 1984, AT&T introduced its new logo to indicate its new "world" view. (Photo courtesy of AT&T Archives.)

in AT&T's logo that was chosen in 1984 for the "new" AT&T. Even though its logo has remained the same, AT&T has evolved and changed considerably since then in what has come to be considered a period of great turmoil for the telecommunication industry and the old Bell family.

Postdivestiture turmoil

In the 15 years since divestiture, AT&T and the Baby Bells have behaved much like children in the sense of demanding that which is forbidden. First, it was AT&T that so strongly wanted to be in the computer business after years of being restricted to the provision of only telecommunication products and services. Then, it became the Baby Bells who clamored to be allowed into manufacturing and long-distance service. Judge Greene, who continued to oversee the enforcement of the MFJ, refused to allow the Baby Bells into these businesses. The Baby Bells instead bypassed him by lobbying to the U.S. Congress for relief from the restrictions of the MFJ—and ultimately they prevailed.

The Baby Bells became obsessed with the concept of the convergence of entertainment with telecommunication through the use of a single broadband medium to the home. Cable television, telephone service, and data communication would all be provided over a single broadband medium, either optical fiber or coaxial cable. In pursuing this dream, some Baby Bells formed close alliances with cable companies, and Bell Atlantic almost purchased the cable company TCI in 1994 until sound business sense finally prevailed. US WEST purchased a 25 percent ownership in Time Warner Entertainment, and many of the Baby Bells—along with a number of media companies—initiated costly trials of interactive video services to the home. By the end of the 1990s, none of this worked out for anyone. In 1998, for example, US WEST divested its cable ventures into the creation of MediaOne.

A major factor behind the MFJ was AT&T's desire to be free of the restrictions of the 1956 Final Judgment. AT&T had become convinced that telecommunications and computers were converging and that it had to have a strong presence in the computer business to survive. A similar belief in convergence led IBM into acquiring Satellite Business Systems—a big mistake for IBM. AT&T initially attempted to develop its own line of personal computers around its own proprietary chips. This failed, and AT&T then acquired a part ownership in the Italian firm Olivetti in 1985. Olivetti knew little of the computer business, and AT&T sold its ownership eight years later—actually at a profit.

When it became clear that the Olivetti connection was not going to work, AT&T sought another partner as a way to strengthen its computer business, and in 1991 purchased NCR for $7.5 billion. AT&T continued to be convinced that its mission was "… to be the world leader in information technology—a technology that inescapably involves the best of both computer and communication technology." [AT&T's Richard Bodman, writing in *The New York Times*, April 14, 1991.] But these synergies failed to materialize, and AT&T finally had to accept the fact that remaining in manufacturing was too costly in general and was misfocusing its attention from its core telecommunication services. Accordingly, AT&T in 1996 divested its manufacturing arm, Western Electric, to create a new manufacturer, Lucent Technologies, Incorporated. At the same time, AT&T also divested NCR. The action was called a trivestiture since it split AT&T into three components. But the biggest event of 1996 was the signing into law of new telecommunications legislation, which opened the door to the Baby Bells to enter the long-distance business by freeing them from the restrictions of the MFJ.

The Telecommunications Act of 1996

The Telecommunications Act of 1996 was overwhelmingly passed by the U.S. Congress in February 1996, and then signed into law by President Clinton on February 8, 1996. The act reforms telecommunications by allowing the Baby Bells into long distance and video, deregulating cable TV, liberalizing the ownership of broadcast services, and eliminating the restrictions of past Consent Decrees involving AT&T, GTE, and McCaw.

The intent of the portions of the act dealing with telephone service was to substitute competition for regulation under the premises that telephone companies would provide video in competition with cable companies, cable companies would provide local telephone service, and entry into long distance by the Baby Bells would lower long-distance rates. The FCC was given the task of determining the specific orders to implement the overall intentions of the Act.

The act allowed the Baby Bells into the provision of long-distance service outside their serving areas with no restrictions. However, because of their monopoly over local access, restrictions were placed on the provision of long-distance service within their serving area. The condition was that the Baby Bells open local service to competition. In August 1996, the FCC mandated a number of access points in the local exchange and ordered that these interconnection points be made available to any and all telecommunication providers. The FCC also attempted to specify the method for determining pricing for the opened local exchange along with default prices.

The Congress was concerned that universal service would not be served with competition. Accordingly, in August 1997, the FCC issued the orders for the creation of a Universal Service Fund to be contributed to by all providers of telecommunication services. The fund gives financial support to schools, libraries, rural health care providers, high-cost and rural areas, and low-income subscribers. The Universal Service Fund was estimated to be nearly $5 billion, and since its costs are ultimately passed on to telephone users, it is, in effect, a tax of roughly 5 percent on the provision of telecommunication. The intent of the FCC was to make specific all the inherent past subsidies of universal service and to create a support mechanism to which all providers of telecommunication service would contribute.

In its orders issued in 1997, the FCC also defined universal service. Universal service was defined to include voice-grade access to the public switched telephone network; dual-tone multifrequency signaling or a digital functional equivalent; single-party service; toll limitation services for qualifying low-income consumers; and access to such services as operator, directory assistance, and emergency 911. The FCC also attempted to reform access charges, and this topic is treated in the chapter that follows.

Merger mania—the turmoil continues

Confusion, conflict, contention, chaos—all beginning with "c" seem to characterize competition in telecommunication. The turmoil in telecommunication has continued throughout the 1990s, and actually seems to have worsened. Pacific Telesis became convinced that its future was in wireless and other Information-Age businesses, and accordingly contemplated divesting its local telephone service business. Sense prevailed, and instead, Pacific divested its wireless business with the creation of Air-Touch in 1994. Then, in 1997, Pacific Telesis was acquired by SBC Communications in what became one of the first moves in the reformation of the old Bell family. Also, in 1997, Bell Atlantic acquired NYNEX. The seven Baby Bells were now only five.

In 1997, it seemed that nearly everyone wanted to buy MCI. First, it was British Telecom, then GTE, and finally a deal was finalized in 1998 with WorldCom as the new owner. What was particularly baffling about GTE's attempt to acquire MCI was that GTE was in the long-distance business and then abandoned it. In 1986, GTE acquired a 50 percent interest in Sprint, and then sold this interest in 1992. The summer of 1997 was particularly ripe with craziness. Not only was everyone chasing MCI, but rumors surfaced that AT&T wanted to buy SBC Communications. Nothing materialized, but after a management crisis, AT&T finally did acquire a new chairman in November 1997. AT&T sold its undersea cable operations to Tyco in 1997 for $850 million.

In 1998, AT&T acquired Teleport and announced its planned acquisition of the cable company Tele-Communications Inc. (TCI)—the same TCI that Bell Atlantic almost bought in 1994—for a $48 billion deal. A condition to the WorldCom-MCI merger was that MCI divest itself of its Internet backbone network, which it did by selling this business to the British firm Cable & Wireless for $1.75 billion. SBC Communications acquired Southern New England Telephone (SNET) in 1998 and made a deal to acquire Ameritech. Soon, there will be only four Baby Bells! And Bell Atlantic just announcved its intention to acquire GTE. Will merger-mainia ever end?

In 1998, as the third edition of this book is being written, it seems as if the old Bell System is re-forming. The history of the Bell family, depicted in Figure 12.5, shows a tremendous fragmentation in 1984, and then further fragmentation of AT&T as manufacturing is finally divested. The Baby Bells, though, have been coming back together. It is interesting to speculate what would have happened in 1984 if AT&T had retained its ownership of the local Bell telephone companies and instead divested its long-distance and manufacturing businesses.

An uncertain future

This story of the history of telecommunications in the United States, with an emphasis on the Bell System, must come to an abrupt end. While this book is being written in 1998, it is not clear where telecommunication is going in the United States. A stable point has not been reached. AT&T is moving into cable television. The Baby Bells are merging. Merger mania seems to be afflicting telecommunication and Wall Street. How can one maintain a perspective in the face of all this turmoil? That is the topic of the last, and following, chapter of this book. The answers will attempt to

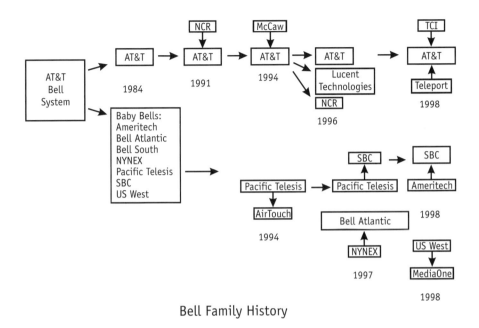

Bell Family History

Figure 12.5 The Bell family has evolved considerably after the breakup of 1984. It most likely will continue to do so, perhaps even re-forming, but finally divorced of manufacturing.

draw lessons from the patterns of the history that were just described in the preceding material.

Selected bibliography

Brooks, J., *Telephone: The First Hundred Years*, New York, NY: Harper & Row, Publishers, 1976.

Goulden, J. C., *Monopoly*, New York, NY: Pocket Books, 1970.

Kinsley, M., *Outer Space and Inner Sanctums*, New York, NY: John Wiley & Sons, 1976.

Temin, P., and L. Galambos, *The Fall of the Bell System: A Study in Prices and Politics*, Cambridge, MA: Cambridge University Press, 1987.

von Auw, A., *Heritage and Destiny*, New York, NY: Praeger, 1983.

13

A Perspective on Telecommunications

Introduction

I cannot let our story of telephone communication end without giving some final thoughts and perspectives on its future—as I have done so in each edition of this book.

In writing this third revision, I reread the earlier perspectives to be able to summarize and share trends and conclusions with you, to delete outdated details, and to respond to more recent events. This chapter should be read along with the preceding chapter on the history of the Bell System, although the conclusions of some of that history are reviewed here, but from a more personal and opinionated perspective.

Vail's invention

The Bell System was invented over 100 years ago by Theodore N. Vail, who was president of the American Telephone and Telegraph Company from 1907 to 1919. Telecommunication was a hodge-podge of competing companies during the early days of telephony, particularly following the expiration of Alexander Graham Bell's patents. Vail believed that all this competition did not make much sense and rationalized this belief by claiming that the public would be served best if telephony service was provided by a single company on a nationwide scale. The concept of natural monopoly for telephone service was thus promulgated, but Vail's true genius lay in his invention and endorsement of public regulation of the monopoly to ensure that the public interest was best served.

Both Bell and Vail believed in total end-to-end provision of telephone service, including the telephone instruments, the wiring in homes and businesses, and the transmission and switching equipment and systems that interconnected the telephones. Bell and Vail had the vision to realize that profits would ultimately be greatest in the long term if everyone could afford to have telephone service. Thus, the concept of universal service was born.

The goal of universally affordable telephone service was achieved decades ago. This goal was achieved partially by using such services as long-distance and touch-tone to subsidize local service. These subsidizations later created opportunities for competition in the sale of telephones and in the provision of long-distance telecommunication. The insistence by the Bell System on the total end-to-end control of telephone service became less and less tenable. Consumers could purchase and connect their own telephones to the local line. A number of firms were supplying long-distance service in open competition with AT&T. The Bell monopoly was crumbling, and its final official fragmentation occurred on January 1, 1984, under the terms of the Modified Final Judgment reached between AT&T and the U.S. Department of Justice in 1982.

The times since the Bell breakup have been very tumultuous. AT&T's share of the long-distance market shrunk considerably during the 1990s. The Baby Bells grew during this period and actually started to merge as if the old Bell System were about to rise again. As the millennium approaches, the future structure of the telecommunication industry is uncertain.

This chapter explores the causes and ramifications of this fragmentation of the Bell System and its aftermath. Technological, historical, financial, and my personal views are all integrated together in this chapter to give the reader some perspectives on the past, present, and future of telecommunication. Many crucial lessons are learned from the past that are particularly relevant to the future of telecommunication.

Competition comes to telecommunication

The management of the old Bell System had come to believe that AT&T had a government-endorsed right to monopolize telecommunication in the United States. Perhaps this belief was well founded, since on the many occasions that the government attempted to break up the Bell System, the various concessions made by AT&T were able to keep the Bell System intact, and the government always abandoned its cases.

A serious attempt to dismantle the Bell System was initiated in 1949 by the Justice Department. That action was terminated in 1956 by the terms of the Final Judgment. The next attempt to dismantle the Bell System came 20 years later in 1974. That action was settled in 1982 under the terms of a modification of the 1956 Final Judgment. This Modified Final Judgment (MFJ) finally resulted in the dismantling of the Bell System that the government had sought to achieve for nearly a century. Under the terms of the MFJ, the Bell operating companies were divested from AT&T; AT&T retained Western Electric and its Long Lines Division. This divestiture occurred on January 1, 1984. The Bell System was no more.

Such a drastic action was the result of many events and threats, but perhaps one factor overshadowed all the various events and threats. That factor was an attitude problem on the part of AT&T management. For decades, Bell System management believed that service to the public was the reason for its existence and that they alone knew best what was in the public's best interest. Management, in concert with state and federal regulators, made decisions about what services would be overpriced in order to subsidize other services, supposedly in the best interest of the public. This type of thinking led to a corporate arrogance and intransigence that did not understand or show open-mindedness to any changes or lessening of AT&T's monopolistic control of telecommunication in the United States. Thus, even the smallest threat to the status quo of the AT&T monopoly was met by an overwhelming, defensive response from the Bell System—and, indeed, a whole series of such threats occurred.

Today, all this is quite different. The local telephone companies are barred from providing telephone instruments as part of telephone service. Any customer-owned equipment can be connected to the line, although it must be type-certified by the FCC. This was facilitated by the development of the modular plug and socket by Bell Labs' engineers to make it easier for telephone-company installers to connect and disconnect telephones. The same modular plug made telephones as easy to install as electric appliances—all you need do is simply plug them in. The Bell monopoly on telephone instruments had finally died.

The collapse of the Bell System

The settlement reached by AT&T and the Justice Department freed AT&T from all the restrictions of the 1956 Final Judgment. In return, the local telephone companies were divested from AT&T and were barred from manufacturing and from providing long-distance service. Intrastate toll service was given to AT&T. In terms of its new vision of the future, AT&T clearly seemed to have won. Furthermore, AT&T had nothing to fear from any competition from the local companies, since they were barred from manufacturing and long-distance service—the "new" AT&T's two major business areas. It was almost as if AT&T itself wrote the Modification of the Final Judgment (MFJ) to its best interests.

A large component of AT&T's business was, and still is, the provision of long-distance service. Although AT&T had hoped otherwise, that business was still subject to regulation after divestiture, and a host of competitors competed vigorously with AT&T.

There is an interesting way to view divestiture. In 1988, the total value of property, plant, and equipment (including accumulated depreciation) owned by the seven regional holding companies (RHCs or Baby Bells) was $186.4 billion vs. AT&T's total value of only $40.9 billion.

Clearly, the operating companies formed the bulk of the Bell System's assets, and the Bell operating companies (BOCs or local telcos) were the Bell System. Viewed this way, AT&T in effect had divested itself of the Bell System!

Divestiture was a complex process, and it has taken some time for all the dust to settle. We shall see later in this chapter who really won—and lost—from the drastic restructuring of the telecommunication industry in the United States.

It is interesting to contemplate alternative actions that might have occurred in 1984. AT&T might have retained its ownership of the local Bell telephone companies and divested manufacturing and long distance. If this had happened, AT&T in 1997 would have had revenues of $102 billion and a net income of $10.7 billion. Instead, AT&T actually had revenues in 1997 of $51 billion and a net income of $4.6 billion. AT&T's children have indeed done well since leaving their former Bell home.

Since divestiture, AT&T itself has undergone a series of reorganizations, acquisitions, and divestitures as it attempts to redefine its business and markets. The frequency of these activities, along with their complexity, makes it a challenge to comprehend AT&T's strategic direction at any particular moment. But perhaps learning how to cope with uncertainty and constant change is the challenge of the day, and not only for AT&T.

Access charges

To provide long-distance service, carriers need access to the local network both to complete a long-distance call in a distant city and to give access to the calling party. Long-distance carriers pay the local telephone companies for this access through the imposition of access charges. In 1988, the RHCs received 27 percent of their revenue in the form of access charges paid by long-distance carriers such as AT&T. In that same year, AT&T paid 48 percent of its long-distance receipts to local telephone companies for access to their local facilities. In 1997, 24 percent of RHC revenue was obtained from access charges, and AT&T paid 32 percent of its total revenue for access. Although the impact of access charges has decreased somewhat over these 10 years, access charges are still the glue that continues to maintain an intimate financial relationship between AT&T (and other long-distance companies) and the BOCs.

Access charges are not new, although in the past they were called separations. AT&T's Long Lines Department returned a large proportion of the charges collected by the local telephone companies back to them as reimbursement for the use of their local facilities in completing long-distance calls. Nearly 60 cents of every dollar in toll charges was returned to the local company and, in theory, helped subsidize local service. The local companies encountered costs in supplying billing, switching, common local transmission, and other services associated with long-distance service. These costs are difficult to determine, and thus I am not sure whether the 60 cents on the dollar was really sufficient payment to the local telephone company. However, the accepted wisdom is that toll service subsidized local service. The returned money was a sizable source of revenue for the local companies and perhaps could be viewed as the financial glue that kept the Bell companies in the AT&T empire.

In reality, the only reasons MCI was able initially to compete with AT&T were (1) that MCI did not subsidize local service by paying large fees to the local telephone companies for access to the local network and (2) that MCI did not have to provide service to less lucrative cities and rural areas. AT&T's arguments against MCI at the time were legitimate, but AT&T was viewed as a "nasty monopoly," simply protecting its position. AT&T turned to the political arena to obtain relief.

With the 1984 divestiture, the subsidy—if indeed it was one—still had to be paid or the local companies would lose a large source of income and local rates would need to be increased accordingly. The subsidy became an access charge, initially paid solely by long-distance carriers. To decrease the access charges paid by these carriers, the FCC imposed a flat increase in local rates to be paid by all telephone subscribers, and this new fee was also called an access charge. The rationale for the subscriber access charge was that if carriers were being charged too much for access, they would bypass the facilities of the local telephone company and connect directly to large business customers. This would deny revenue to the local company, and local rates would increase.

The Telecommunications Act of 1996 attempted to decrease the per-minute access charges by offsetting them with increases in the flat subscriber line charges and with new flat charges to be paid by the long-distance companies. The final effect of these reforms was neutral for the local companies.

Access charges continue to be a serious problem as the 1990s close. Local calls to Internet access providers do not incur any per-minute access charges. However, calls to be connected to a long-distance provider do incur per-minute access charges. This creates a false incentive to use the Internet for telephone service since about 4 cents a minute would be saved by bypassing the access charges. Furthermore, the long holding times and short bursts of data traffic of Internet calls ties up the local switch in mostly switching idle silence. The obvious solution is the imposition of local usage sensitive pricing for all calls, perhaps something in the order of a penny a minute.

Another solution is the elimination of all usage-sensitive access charges at the local level. The impact of this on the local telephone companies would be reduced by allowing them to provide intrastate long-distance service and to then use these new profits to reduce the need for access charges. The BOCs provided intrastate long-distance service before divestiture, but the stipulations of the MFJ allowed AT&T to take away this very profitable business from the BOCs.

Access charges continue to be a controversial topic and are one example of the lack of foresight of all the consequences of competition in telecommunication and of divestiture.

Subsidies

The Telecommunications Act of 1996 continues to promote universal service, even though nearly everyone who wants one has telephone service in the United States. In the name of universal service, the act creates subsidies for schools and libraries, high-cost areas, and low-income consumers. The subsidies for schools, libraries, and rural healthcare providers are intended to promote access to electronic information. The subsidies for high-cost, rural, and other areas are intended to eliminate rate inequities so that everyone pays the same for telecommunication service. This all sounds much like a socialistic mentality, which is perplexing since the stated grand purpose of the act is to enhance competition. The universal service subsidies resulted in 1998 in a "tax" of about 5 percent of the revenues of all telecommunication service providers, although this tax might rise over time.

The entire mentality of subsidies and universal service is outdated and is inconsistent with a free marketplace and competition. The real problem was created decades ago with the subsidization of local service by long-distance service. Today's version of this inequity is access charges, which have been extended to create explicit subsidies for various social purposes. Social welfare indeed has its place, but should not be funded by charges on specific industries. It would not make sense to charge fast-food establishments some extra tax that would then subsidize the hungry. So, too, telecommunication service providers should not be taxed explicitly to subsidize phone service to the needy or to schools and libraries or to rural areas. This subsidization problem also occurs with 911 service, which is funded mostly by surcharges on phone bills.

Subsidies once initiated are difficult to halt. Those benefiting become a strong political voice continuing to beg to be fed. Even in those cases where help has not been requested, who will not refuse a handout from the government when it is offered? All this distorts market forces, and these distortions are not beneficial to real competition. The assistance for schools and libraries was lobbied for not by schools and libraries, but mostly by the Baby Bells. Not surprisingly, the Baby Bells are the ones who ultimately receive the aid given to the schools and libraries, at the expense of inter-LATA service providers.

The development of the Internet was strongly financed by the federal government. One could argue that this is a fine example of government subsidization that has benefited the public. However, the telecommunication industry in the United States is very healthy and profitable. It does not need government funding. The government, in effect, nationalized the Internet—a public packet-switched data network. The problem is that once the government embraces anything, it is nearly impossible to extricate it. In the mid 1990s, the National Science Foundation halted its support of the Internet backbone, but then initiated funding of the development of an advanced packet network for its research community. This advanced network will probably also evolve into a public data network. I think it is time to privatize data communication in the United States and cease any and all government development of data networks.

The RHCs

Prior to divestiture, the stock of the Bell operating companies (BOCs) was owned and held by AT&T. During the process of divestiture, AT&T created seven regional holding companies (RHCs) and transferred the ownership of the BOC stock to the RHCs. Stock in the RHCs was distributed to AT&T shareowners in proportion to their AT&T holdings. One rationale for the creation of the RHCs was AT&T's belief that some of the BOCs were too small to be financially viable. Combining BOCs together would create seven larger and stronger companies. However, the regional holding company is simply a smaller replacement of the AT&T bureaucracy. The BOCs never achieved true independence—they now are owned by RHCs.

The provision of local service is the main business of the BOCs and thus their owners, the RHCs. Many of the RHCs entered other businesses, as allowed initially by Judge Greene. However, most of the revenue and profits of the RHCs still comes from the BOCs and the provision of local telephone service. We shall see later in this chapter that a large portion of that revenue comes mostly from AT&T in the form of access charges. AT&T and the BOCs are still pretty much welded together.

Immediately after divestiture, the RHCs attempted to enter new business areas. Judge Harold Greene, who presided over the divestiture case and continued to oversee its provisions until the 1886 Telecom Act, put a halt to some of these planned ventures and stressed that the major business of the RHCs was the provision of local telephone service within their assigned local access and transport areas. The fear was that the RHCs would somehow subsidize the new business ventures from the profits of their regulated telephone business. The RHCs continued to hope for relief from these restrictions, and finally achieved relief with the provisions of the Telecommunications Act of 1996, which allowed them into manufacturing. However, why anyone would want to be a manufacturer in today's information and service economy eludes me.

The antitrust case whose settlement resulted in divestiture was primarily against AT&T, Western Electric, and Bell Labs. Nevertheless, it was the BOCs that were "punished" by being divested from AT&T, and AT&T was "rewarded" by being freed from the restrictions of the 1956 case. The answer to who was really punished and who was really rewarded is best determined by examining the financial effects of divestiture on AT&T and the RHCs.

Financial effects of divestiture

In 1982, the year before divestiture, AT&T had a net income of $7.3 billion, or a net income equal to 11.1percent of revenue (commonly know as after-tax profit margin). In 1984, the first full year of operation after divestiture, AT&T's net income dropped to $1.4 billion with a profit margin of only 2.6 percent. In reporting its income after divestiture, AT&T routinely excludes the access charges it pays to the local telephone companies. Since these access charges are collected by AT&T and represent a cost of doing business, I have included them in my calculations of AT&T's revenues. In 1984, the RHCs as a group had a net income of $6.8 billion and a profit margin of 11.8 percent. Clearly, the BOCs were the financial strength of the Bell System, and AT&T was crippled financially as a result of their divestiture.

Figures 13.1 and 13.2 show graphically the financial results of divestiture in terms of net income and profit margin in detail from 1982 to 1989, and, in summary, for 1997. The dip in AT&T's financials in 1986 was caused by the setting aside of a large sum of money to finance early retirements and to stimulate resignations in order to reduce the size of AT&T's labor force. The 1988 loss was the result of accelerated depreciation and the replacement of analog multiplexing facilities in the AT&T long-distance network. The trend, ignoring these two dips, in AT&T's financials was modest growth in net income at a compound annual growth rate of 14 percent over the five-year period of 1984 to 1989.

The financial effects of divestiture clearly show the considerable health of the BOCs and RHCs. The surprise was the comparatively poor performance of AT&T. We shall later examine some possible reasons for this. But overall, the breakup of 1984 has been a financial success.

In 1997, the former Bell family consisting of the five Baby Bells, AT&T, and Lucent Technologies had a combined revenue of $222.6 billion and a combined net income of $15.8 billion. The overall profit margin in 1997 for the former Bell family was 7.1 percent. Thus, in the 14 years since the breakup, the former Bell family has more than tripled its

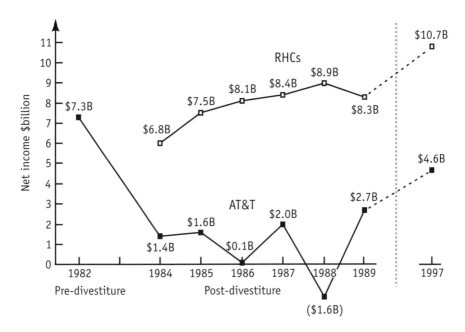

Figure 13.1 The net income of the seven regional holding companies (RHCs) climbed steadily in the first six years after divestiture. A small decrease in 1989 was a result of an earnings decrease at NYNEX. AT&T's earnings varied greatly during this six-year period. The 1986 dip was a result of placing aside funds to stimulate employees to resign or retire early. The 1988 dip was a result of accelerated digitization of the AT&T network.

revenues and more than doubled its net income. The overall profitability has decreased, though, probably because of competitive pressures on prices in long-distance services and manufacturing and because of regulatory scrutiny by some state commissions.

Research and development

In this book, we have seen the importance of science and technology and of research and development (R&D) in the provision of telephone service. In the days of the old Bell System, much of that R&D was performed at Bell Laboratories. Bell Labs' engineers designed the equipment and systems that were manufactured by Western Electric and used by the Bell

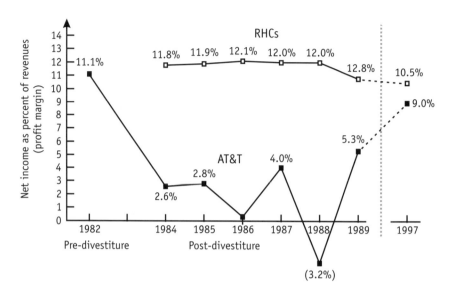

Figure 13.2   The profit margins of the seven regional holding companies (RHCs) and of AT&T are plotted for the fist six years after divestiture. Discounting dips in 1986 and 1988, AT&T's overall trend shows growth in profit margins for the first six years after divestiture, although the figures are at least half those of the seven RHCs. The data for 1997 shows AT&T profitability that finally nearly equals that of the RHCs.

operating companies and Long Lines. About 10 percent of Bell Labs' resources were devoted to performing basic research in areas deemed applicable to telecommunication, and much of Bell Labs' reputation was based on the accomplishments of its researchers. Indeed, Bell Labs had come to be considered a national R&D resource. Quite understandably, there was concern that divestiture and the restructuring of telecommunication that occurred in 1984 might somehow damage Bell Labs and telecommunication R&D.

A few years after divestiture, I performed a study to assess its effects on Bell R&D. Two years before divestiture, at the end of 1981, AT&T spent $1.6 billion for R&D conducted mostly at Bell Labs, and 24,100 people were employed at Bell Labs. About 1,200 people were involved directly in performing basic research. With divestiture, Bell Labs was required to change its name to indicate its continued affiliation with AT&T, and Bell Labs was renamed AT&T Bell Labs.

AT&T identified the need for the BOCs to have their own central research organization to address common technical and scientific questions. This need was addressed by the creation of a central service organization called Bellcore, fully owned and funded by the seven RHCs to be responsive to their needs and direction. Hence, the postdivestiture R&D picture must combine the AT&T R&D effort with Bellcore. By doing this combining, I found that at the end of 1985 there were 32,000 R&D employees at a combined funding level of $3.2 billion. The number of researchers had likewise increased from a predivestiture number of 1,200 to a postdivestiture combined number of 1,700.

In 1989, AT&T spent nearly $2 billion on product-related R&D. Actual sales of products generated only $11.5 billion, and hence product R&D was more than 17 percent of sales. This is very high, and indicates the high cost of AT&T R&D in this age of intense competition. This R&D was in effect being subsidized by the profits generated by AT&T's very successful long-distance business.

In 1989, the RHCs spent about $1 billion for R&D performed for them by Bellcore. This amounted to less than 1.5 percent of their combined revenues of $77 billion for that year. Given the high technology and complexity of today's telecommunication system coupled with its essential role in our information and service economy, this amount of R&D was perhaps not adequate. But since the RHCs were not involved in any manufacturing activities, perhaps much less R&D was needed for their predominately service-based network business.

In terms of numbers and figures, R&D fared quite well through divestiture. But as the 1990s were entered, new questions about the health of telecommunication R&D arose.

In the early 1990s, the RHCs became disenchanted with the philosophy of a centralized R&D organization and initiated actions to sever their guaranteed support of Bellcore. Bellcore then went through a very rocky period of much uncertainty about its future. In 1997, Bellcore was finally sold by the RHCs to Science Applications International Corporation. In 1997, Bellcore had a budget of $1.1 billion and 5,600 employees, with 300 involved in applied research. Much of its work is still supported by the RHCs.

In 1996, AT&T divested its manufacturing business with the creation of Lucent Technologies. A very large portion of the former AT&T Bell

Labs went along with Lucent and is called Bell Labs. Lucent reminds us all of its heritage of invention by mentioning "Bell Labs Innovations" in its corporate image and advertising. In 1997, Lucent spent $3.0 billion to support research and development at Bell Labs. Bell Labs had about 1,200 permanent people directly involved in performing basic research. Based on these large numbers, it seems that Lucent inherited and supports much of the past research and development tradition of the old Bell Labs.

AT&T maintains its own research and development facility called AT&T Labs. In 1997, it had a total of 2,000 employees, including 400 researchers. I estimated the total research and development budget for AT&T Labs to be very roughly about $0.3 billion.

Thus, the total research and development effort of the fragmented Bell family in 1997 was $4.4 billion, which comprised $0.3 billion for AT&T plus $3.0 billion for Lucent plus $1.1 billion for Bellcore. The total number of people in research and development was 32,600, consisting of 2,000 for AT&T plus 25,000 for Lucent plus 5,600 for Bellcore. Looking only at people working in pure and applied research, the total number of research people was 1,900, consisting of 400 for AT&T plus 1,200 for Lucent Bell Labs plus 300 for Bellcore. Compared to the predivestiture levels, research and development in the old Bell family is larger in terms of the number of people and appears well supported financially. But numbers alone do not tell the whole picture.

Bell Labs of the past was a great research organization. I know this first hand since I worked there for well over a decade in the 1960s and early 1970s, performing research in a number of exciting and novel areas. What contributed to the greatness of Bell Labs as a research institution were such factors as secure, long-term funding; a strong sense of a mission to ensure the long-term future of telecommunication; and the freedom to take risks and to fail.

AT&T and Lucent are not the same as predivestiture AT&T. AT&T and Lucent are now subject to intense competition and revenue fluctuations; secure, long-term funding cannot be assured anymore. AT&T's mission has been changing with the fancies of management; a strong sense of mission is absent. Lucent's mission as a manufacturer of telecommunications products and systems does seem clear and stable. However, both AT&T and Lucent must be concerned about the market success of new products and services; the freedom to take risks is much reduced. The new emphasis clearly is on relevance; research must be more applied and its results must be quickly transferred to technology relevant to existing products, services, and businesses. Yet technology transfer is a tricky business and frequently occurs outside formal organizations.

Since it was AT&T that claimed before divestiture that Bell Labs was a national treasure and resource, the health of postdivestiture research at AT&T Labs, Lucent's Bell Labs, and Bellcore continues to be a national question. The real issue, though, should be research in telecommunication in general. The past situation in which AT&T had been funding basic research performed at Bell Labs that benefited the whole telecommunication industry was untenable in the long run. Somehow, the whole industry must assume more responsibility for funding research. I do not know the solution, but perhaps some combination of the basic research performed at Bellcore, AT&T Labs, and Lucent's Bell Labs would make sense, with broad funding from the whole industry. Basic research needs to be stimulated by real problems, and research performed at universities is usually too far removed from the problems of the real world. But the results of basic research are published and any proprietary ownership is difficult, particularly given the long-term horizons of much basic research.

AT&T: a status report

Before the divestiture of manufacturing, AT&T had three major businesses: sale of services, in particular, telecommunication services such as long-distance telephony; sale of products, such as telephones, switching systems, and PBXs; and rentals. In 1989, these three business segments generated $22.1 billion (net of access charges of $14.9 billion), $11.5 billion, and $2.5 billion, respectively. AT&T did not report the actual profit or loss for each separate business. But by examining the data reported to the FCC by AT&T for its communication service business and by making some assumptions, some net incomes could be extracted and showed that AT&T's product business could have lost as much as $4 billion in 1989. AT&T's service business was the "cash cow" that kept AT&T afloat.

Given the size of these losses, I wondered why AT&T continued the sale of products? A number of possible explanations were possible. One explanation was that my estimations of the losses were incorrect. Another explanation was that AT&T hoped to be able to turn the product business around and make it profitable in the future. Another explanation was that AT&T believed it must be able to provide products as part of a total system solution to the communication needs of its customers. And a last reason was that AT&T needed to shelter the profits of its long-distance business or else it would be subject to greater government scrutiny and regulation.

I concluded that from a purely financial perspective, AT&T should divest or otherwise eliminate the sale of products. This would create a much leaner and more sharply focused AT&T. The AT&T service business would be very free to purchase its equipment and systems from any one of many suppliers. This would stimulate the equipment business in a way similar to what happened when the BOCs were freed by divestiture from purchasing most of their equipment from AT&T.

In 1996, AT&T finally divested manufacturing by creating Lucent Technologies. In its first year, Lucent had a profit of $1 billion. For 1997, Lucent had a profit of $1.5 billion on revenues of $26.4 billion—an after-tax profit margin of about 6percent. Clearly, Lucent is a financial success, which is a surprise to me given the large losses that I had estimated for AT&T's Western Electric manufacturing entity. Lucent is a most impressive financial turnaround, and its stock soared during its first few years. Given Lucent's great success, AT&T must be pleased with the divestiture, but must be asking itself what it was doing wrong when it owned Lucent's previous incarnation, Western Electric. After abandoning manufacturing, AT&T developed a new vision.

In the late 1990s, AT&T became convinced that it must have a presence in local telephone service distribution. In 1997, rumors surfaced that AT&T was contemplating an acquisition of SBC Communications. And in 1998, AT&T announced its planned acquisition of the cable company Tele-Communications Inc. (TCI) for a deal worth $48 billion. The merger triples AT&T's debt and takes on TCI's yearly losses of $0.5 billion. The price paid equates to over $3,000 per cable subscriber. AT&T will need to make a profit of over $50 per month per cable subscriber to recoup its investment in 15 years with a 15 percent return. Clearly, with average monthly cable bills of less than $40, AT&T can never make an acceptable return on its investment. It would appear that AT&T has bet its financial future on the TCI merger—and will perhaps lose it all.

There certainly appears to be much controversy and opinion about AT&T's future. Much change is probably still in store for AT&T.

The Baby Bells: a status report

On the whole, the Baby Bells have been a great success. Their profit margins are stable and are quite good. Productivity measured as the number of access lines per employee continues to grow. In 1997, all of their access lines were served by computer-controlled switching systems. Network usage continues to grow. By any and all measures, the BOCs have been a great success. And yet, the BOCs were not content with what they had.

The BOCs wanted to be able to provide information services and long-distance service, and to be involved with manufacturing. In the late 1980s, the BOCs initiated advertising and lobbying efforts to obtain relief from the restrictions of the Modified Final Judgment so that they could enter these other businesses. The Congress responded with the Telecommunications Act of 1996, which allows the Baby Bells into these areas, but subject to some restrictions.

To some extent, the BOCs have good reason to want to be involved with manufacturing. Closer ties with manufacturers might allow the BOCs to help speed the innovation and introduction of new telecommunication services and products. However, given AT&T's problems with manufacturing and the sale of products, the BOCs would be well advised to avoid any direct involvement in manufacturing. The restrictions of the MFJ against manufacturing actually seem to have saved the BOCs from replicating AT&T's losses in the product area.

The BOCs still provide telephone directories along with their very profitable Yellow-Pages listings and advertising. Won't it be nice to have a shopping service in which you telephone a directory operator who then uses a computer terminal to search a database, and then gives you the names and addresses of the nearest stores along with information about any special sales. Should such a service be supplied by the telephone company or by others? Should the telephone company be forbidden, as it now is, from supplying such a service? A laser disk can easily store all the information in a number of telephone books. Should the telephone company create and distribute such disks? Computerized databases for the home and accessed over the Internet are much the way of the future. Should the telephone companies be allowed to own and operate such databases? What about owning and operating gateways to databases owned and operated by others?

The issues and questions concerning the provision of information services by the BOCs and RHCs are most challenging and interesting. One thing for certain is that many of these new information services are very risky and could be a great opportunity to lose many millions—perhaps even billions—of dollars.

The one business that is not risky is the transmission and switching of the signals that give access to these services. The transmission and switching of signals—the conduit business—is indeed the business of the BOCs. In most cases, the BOCs would probably be well advised to avoid the hype and risk of many of these new information services and instead concentrate on the transmission and switching of signals.

At the Bell breakup in 1984, AT&T took intrastate long-distance service away from the Baby Bells. Much long-distance traffic is intrastate, inter-LATA calls, which could be easily carried by the Baby Bells over their existing network. However, the Telecommunications Act of 1996 allowed the Baby Bells into long distance but subject to the opening of their local business to competition. The conditions imposed for this by the FCC in 1997 were so stringent that no Baby Bell qualified. The Baby Bells attacked the impositions of any conditions as being unfair since GTE was not subject to any and no court had found the Baby Bells guilty of any infractions of the law that would warrant such punishment. There is much logic to this position, and the whole matter was still under discussion during 1998. Clearly, AT&T and the other long-distance companies are fearful of the local marketing power of the Baby Bells should they be allowed back into the intrastate long-distance market—a business they were in before 1984.

As the 20th century comes to a close, the Baby Bells are shrinking in number as they merge with each other. At the Bell breakup, there were seven Baby Bells. In 1998, the seven were already reduced to four with the acquisition of NYNEX by Bell Atlantic and the acquisition of Pacific

Telesis and Ameritech by SBC Communications. In 1998, Bell Atlantic announced its intent to acquire GTE, the largest independent non-Bell phone company. How much longer will US West and Bell South stand alone? Will the merger process among the Baby Bells continue until there is only one mega Bell?

The Baby Bells appear to be following different strategies in their mergers. SBC is acting very much like a stockholding company in its mergers. Pacific Bell, Ameritech, and other acquisitions retain their identities in their local markets, and SBC is mostly an invisible owner. Bell Atlantic, however, strongly exerts its identity. The NYNEX name was totally eliminated after it was acquired by Bell Atlantic.

A final assessment

It is indeed difficult to gaze into a crystal ball and ascertain the ultimate effects of divestiture on telecommunication in the United States. In gazing into the crystal ball in the first edition of this book, I raised the question whether AT&T would become an entrepreneurial company able to develop and manage a broad portfolio of technology businesses. Clearly, that has not happened. Competition with AT&T's products has been very keen, and AT&T's foray into developing its own personal computer was a failure and financial disaster.

In the second edition of this book, I observed that the 1980s saw a great upheaval in telecommunication, and accordingly I wondered what surprises were in store for the 1990s. I wondered whether AT&T would divest its manufacturing and product business? Would the BOCs be allowed to enter manufacturing, and if so, would they fare any better than their past owner? Would the BOCs enter the long-distance business? What would be the performance of Northern Telecom, MCI, Sprint, and all the other firms in the telecommunication industry?

There was a tendency in the past to assume that "telecommunication" was synonymous with AT&T and the Bell System. Divestiture most certainly changed that assumption. There are a number of long-distance carriers from which to choose, but AT&T still remains the dominant carrier, although with a dwindling market share. The BOCs still have a virtual monopoly on local service, but bypass and the use of new technologies are mentioned as possible future threats to their dominant position. Telecommunication has become a partnership involving a variety of competitive firms, which together provide end-to-end service. The challenge is to be sure that this sum of parts continues to equal the quality of service of the past.

Competition in customer-premises equipment has resulted in higher quality telephones at lower prices to the consumer. Competition in long-distance service has resulted in considerable savings to those making such calls. However, the rates for local service have increased greatly. This increase was partially an overreaction to the uncertainties of divestiture and also the need to accelerate the depreciation and replacement of old electromechanical switching equipment. Rates for local service should stabilize and actually should decrease as the efficiencies of the newer switching equipment take effect. Overall, it would appear that the consumer has benefited from competition in telecommunication and from the divestiture of the BOCs from AT&T.

Compared with the rest of the world, the United States has excellent telephone service at the lowest rates. Telephone service in most of the world is still a government monopoly, but that situation is changing as some countries follow the United States in seeking more competition in telecommunication and by privatizing its provision.

The major problem with the present competitive system is that no single supplier is responsible for the total end-to-end provision of telephone service. Decisions must now be made about such items as which company to obtain long-distance service from and whose telephone to purchase. Decisions are hard to make, and thus far, that seems to be the major cost of divestiture and competition.

Clearly, the dust from the explosive turmoil of divestiture and the restructuring of telecommunication in the United States is still in the process of settling. Uncertainties, challenging issues, and intriguing questions abound. It certainly seems that the last decade of the 20th century and the beginning of the 21st century will see many more changes in the structure of the telecommunication industry in the United States. The "last chapter" of the continuing story of the aftermath of restructuring is only beginning to be written.

The last chapter

Our book has told the story mostly of the technology of telephone systems and telecommunication. The technology continues to be indeed a marvel and has greatly benefited us in terms of low cost and useful features. Telephony and telecommunication are the keys to the Information Age and today's service economy. We may take telephone service for granted, but we would be lost without it and our planet would come to a halt—it would be *The Day The Earth Stood Still.*

The telecommunication industry is much more fragmented than ever before, although the mergers of the late 1990s make it appear as if it is coming back together. With fragmentation and many telecommunication suppliers, a sense of unity in a single standardized level of service is even more important than ever. Unfortunately, the situation appears to be drifting toward complexity and lack of ease of use from the perspective of consumers and end users. Constant splitting of area codes, complex voice messaging systems, and lengthy call directories are making the telephone more and more difficult to use. The challenge for the future will be a renewal of Bell's vision for a simple form of human telecommunication.

We have learned in our book how station apparatus, transmission, switching, and signaling work together as a system to deliver telephone service. But there are other parts of the system that we have only touched upon. People and institutions, too, are a part of the system that brings us telephone service. Scientists and engineers contribute the science and technology of telephony. Managers and workers make the service work for us from day to day. And the wondrous visions of Bell, Vail, and others, continue to illuminate the way of the future.

Selected bibliography

Noll, A. M., "Bell System R&D Activities: The Impact of Divestiture," *Telecommunications Policy*, Vol. 11, No. 2, June 1987, pp. 161–178.

Noll, A. M., "The Effects of Divestiture on Telecommunications Research," *Journal of Communication*, Vol. 37, No. 1, Winter 1987, pp. 73–80.

Noll, A. M., "The Future of AT&T Bell Labs and Telecommunications Research," *Telecommunications Policy*, Vol. 15, No. 2, April 1991, pp. 101–105.

Noll, A. M., "A Study of Long-Distance Rates: Divestiture Revisited," *Telecommunications Policy*, Vol. 18, No. 5, 1994, pp. 355–362.

Noll, A. M., "The Costs of Competition: FCC Telecommunication Orders of 1997," *Telecommunications Policy*, Vol. 22, No. 1, 1998, pp. 47–56.

Glossary

THIS GLOSSARY DEFINES some basic terms and concepts that are used in the book but are not defined in detail as they appear.

amplifier   An electronic device that makes a signal larger in amplitude. The gain of an amplifier is a measure of how much bigger the output is compared to the input. The gain is usually measured in decibels, or dB.

amplitude modulation (AM)   A technique for varying the maximum amplitude of a high-frequency tone, or sine wave, called the carrier, in synchrony with the instantaneous variations of the amplitude of an input signal, such as speech or music. The effect of amplitude modulation is to translate, or shift, the spectrum of the input signal to a new frequency range. The translated spectrum consists of two portions, called sidebands. One is an exact replica of the spectrum of the input signal and is located above the carrier frequency. This portion is called the upper sideband. The second portion is located below the carrier frequency and is an exact mirror image of the spectrum of the input signal; it is called the lower sideband. Because the same information is contained in both sidebands, sometimes only a single sideband is transmitted—a technique called single-sideband (SSB) transmission. Also, the carrier is sometimes suppressed because it too conveys no information; this is called suppressed carrier (SC) transmission.

bandwidth   The width of the frequencies that comprises a signal or that is allowed to pass through a communication channel. According to the French mathematician Fourier, any complex waveform can be decomposed into the sum of pure tones, or sine waves, at different frequencies and with different maximum amplitudes and phases. A graphical plot of the frequencies that comprise a signal, or that are passed by a communication channel, is called the spectrum.

bps   The abbreviation for bits per second. A bit represents a single binary digit and is either a 0 or a 1.

capacitor   An element used in electrical circuitry, consisting of two conducting surfaces, or plates, separated by an electric insulator. A capacitor blocks the flow of direct current (dc) but allows alternating, or changing, currents (ac) to flow. The higher the frequency of the alternating current, the less opposition encountered to its flow across the capacitor. A capacitor has a capacitance associated with it, according to its ability to block lower frequencies. Capacitance is measured in the unit of farads (F).

current   A measure of the number of electrons flowing per second in an electric circuit. The unit of current is the ampere, abbreviated A.

dB   The abbreviation for decibel, which is a measure of comparison between two quantities. The decibel is the logarithm of the ratio of the two quantities. If the quantities being compared are electric powers, then a change of 3 dB means that one power is twice as large as the other.

dc   The abbreviation for direct current. Direct current is a constant, unchanging electric current that does not vary with time and that flows in one direction. A battery generates a direct current.

diode   An element used in electrical circuitry that passes current flowing in one direction but blocks current attempting to flow in the opposite direction. Diodes are usually constructed from a junction of semiconducting material.

duplex (full, half)   A form of two-way communication. With half-duplex communication, only one party can communicate in one direction at a time. With full-duplex communication, each party can communicate simultaneously in each direction. The push-to-talk system of two-way radio communication is an example of a half-duplex communication system. The telephone system is mostly full-duplex. Communication that is strictly one way, such as broadcast television, is called simplex communication.

filter   Electrical and electronic circuitry designed deliberately to pass or to block certain frequency components in a signal. A filter that passes high frequencies is called a highpass filter (HPF). A filter that passes only low frequencies is called a lowpass filter (LPF).

frequency modulation (FM)   A technique for continuously changing, or modulating, the frequency of a high-frequency tone, or sine wave, in synchrony with the instantaneous variations in amplitude of an input signal, such as speech or music. Frequency modulation translates and also expands the spectrum of the input signal. An FM signal has greater noise immunity than an AM signal, but this increased noise immunity is obtained at the expense of greater bandwidth.

Frequency shift keying (FSK)   A technique for transmitting data in which the frequency of the carrier wave is changed form one value for a binary 0 to another value for a binary 1. Frequency shift keying is a form of frequency modulation. The term "keying" comes from its initial use in the days of telegraph keys.

Hz   The abbreviation for hertz, which is a unit of measurement of the frequency of a time-varying waveform. The frequency of a periodic waveform that has a basic shape that continuously repeats is the number of full repetitions, or cycles, that occurs in one second. High frequencies are usually written using Greek abbreviations. A frequency of 1,000 Hz is written as 1 kHz, where "k" stands for kilo. A frequency of 1,000,000 is written as 1 MHz, where "M" stands for mega. A frequency of 1,000,000,000 is written as 1 GHz, where "G" stands for giga.

impedance   The frequency-dependent opposition to current in an electric circuit. Impedance is measured in the units of ohms (Ω).

inductance   A measure of the ability of a coil of wire, called an inductor, to block high-frequency signals from flowing through it. Inductance is measured in the units of henrys (H).

mA   The abbreviation for milliamperes. An ampere is a measure of the current flowing in an electric circuit. Current is a measure of the number of electrons flowing per second in the circuit.

multiplexing   The combining together of a number of signals or circuits to share a single transmission medium.

ohm   The unit of measurement for the opposition to the flow of current in an electric circuit. The ohm is denoted by the Greek letter omega (Ω).

Ohm's Law   The relationship between the electromotive force E, electric current I, and resistance R in an electric circuit. Ohm's Law is $E = I R$. This equation tells us that the electric current increases proportionately with the electromotive force.

phase shift keying (PSK)   A technique for transmitting data signal in which the phase of the carrier wave indicates whether a binary 0 or a binary 1 is being sent. With differential phase shift keying (DPSK), a change of phase indicates the binary information.

resistance   A measure of the opposition to the flow of direct current in an electric circuit. Resistance is measured in ohms (Ω).

rms   The abbreviation for root mean square. The rms value of a time-varying waveform is the effective value of the waveform in terms of a constant, nonvarying waveform that would generate the same amount of energy. The rms value is calculated as the square root of the mean value of the waveform squared at each instant of time.

sine wave   A pure tone consisting of a single frequency. A sine wave is characterized by its maximum amplitude, frequency, and phase. The phase is a measure of when the sine wave starts. A sine wave is a smoothly time-varying waveform that alternates in polarity, or direction of flow (either positive or negative).

transformer   An electric device consisting of two coils of wire that are closely coupled magnetically with each other. A changing signal flowing in the input coil, or primary, induces a similarly shaped signal in the output coil, or secondary. The voltage amplitude of the output signal can be made larger or smaller than the input signal, depending on the ratio of the number of turns of the primary compared to the secondary. Sometimes there are two secondary coils obtained by connecting a wire, called a tap, at the center of the secondary coil; this is called a center-tapped secondary.

volts   A measure of the electron-moving, or electromotive, force (emf) that causes electric current to flow in an electric circuit. Volts are abbreviated V.

watts   A measure of the electric power of a signal, abbreviated W.

Bibliography

Blyth, W. J.and M. M. Blyth, *Telecommunications: Concepts, Development, and Management*, Indianapolis, IN: Bobbs-Merrill, 1985.

Cannon, Don L. and Gerald Luecki, *Understanding Communications*, Dallas, TX: Texas Instruments, Inc., 1980.

Fihe, J. L. and G.E. Friend, *Understanding Telephone Electronics*, Dallas, TX: Texas Instruments, Inc., 1983.

Engineering and Operations in the Bell System (Second Edition), Murray Hill, NJ: AT&T Bell Laboratories, 1983.

Freeman, R. L., *Telecommunication System Engineering*, New York, NY: John Wiley and Sons, 1980.

Gibson, J. D., *The Communications Handbook*, Boca Raton, FL: CRC Press, Inc., 1997.

Gurrie, M. L. and P. J. O'Connor, *Voice/Data Telecommunications Systems*, Englewood Cliffs, NJ: Prentice-Hall, Inc., 1986.

Martin, J., *Future Developments in Telecommunications (Second Edition)*, Englewood Cliffs, NJ: Prentice-Hall, Inc., 1977.

Martin, J., *Telecommunications and the Computer (Second Edition)*, Englewood Cliffs, NJ: Prentice-Hall, Inc., 1976.

Noll, A. M., *Introduction to Telecommunication Electronics (Second Edition)*, Norwood, MA: Artech House, Inc., 1995.

Parker, S. P. (Editor in Chief), *Communications Source Book*, New York, NY: McGraw-Hill Book Company, 1987.

Pierce, J. R. and A. M. Noll, *Signals: The Science of Telecommunications*, New York, NY: Scientific American Library, 1990.

In addition to the above books, many trade magazines treat telecommunication technology and its business aspects. Some are: *Data Communications*, *IEEE Spectrum*, *Telecommunications*, *Telephony*, *tele.com*.

About the Author

A. Michael Noll is a professor and former dean at the Annenberg School for Communication at the University of Southern California. Professor Noll is a regular contributor of opinion pieces to newspapers and periodicals, has published over seventy-five professional papers, and has been granted six patents. One of his early patents for research performed at Bell Labs covers the basic concepts of today's virtual reality. He has worked for nearly fifteen years performing basic research at Bell Labs in Murray Hill, NJ, in such areas as the effects of media on interpersonal communication, speech signaling processing, aesthetics, and raster-scan computer graphics. Professor Noll is one of the earliest pioneers in the use of digital computers in the visual arts.

Professor Noll worked in the AT&T Consumer products and marketing Department where he performed technical evaluations and identified opportunities for new products and services. In the early 1970s, Noll was on the staff of the President's Science Advisor at the White House. He is a Senior Affiliated Research Fellow at the Columbia Institute for Tele-

Information at Columbia University's Business School and has been an adjunct faculty member of the Interactive Telecommunications Program at New York University's Tisch School of the Arts.

Professor Noll has a Ph.D. in Electrical Engineering from the Polytechnic Institute of Brooklyn, an M.E.E. from New York University, and a B.S.E.E. from Newark College of Engineering.

Index

Understanding Networking Technology: Concepts, Terms, and Trends, Second Edition, Mark Norris

Understanding Token Ring: Protocols and Standards, James T. Carlo, Robert D. Love, Michael S. Siegel, and Kenneth T. Wilson

Videoconferencing and Videotelephony: Technology and Standards, Second Edition, Richard Schaphorst

Visual Telephony, Edward A. Daly and Kathleen J. Hansell

Wide-Area Data Network Performance Engineering, Robert G. Cole and Ravi Ramaswamy

Winning Telco Customers Using Marketing Databases, Rob Mattison

World-Class Telecommunications Service Development, Ellen P. Ward

For further information on these and other Artech House titles, including previously considered out-of-print books now available through our In-Print-Forever® (IPF®) program, contact:

Artech House
685 Canton Street
Norwood, MA 02062
Phone: 781-769-9750
Fax: 781-769-6334
e-mail: artech@artechhouse.com

Artech House
46 Gillingham Street
London SW1V 1AH UK
Phone: +44 (0)20 7596-8750
Fax: +44 (0)20 7630-0166
e-mail: artech-uk@artechhouse.com

Find us on the World Wide Web at:
www.artechhouse.com